Receptors and Recognition

General Editors: P. Cuatrecasas and M. F. Greaves

About the series

Cellular recognition — the process by which cells interact with, and respond to, molecular signals in their environment — plays a crucial role in virtually all important biological functions. These encompass fertilization, infectious interactions, embryonic development, the activity of the nervous system, the regulation of growth and metabolism by hormones and the immune response to foreign antigens. Although our knowledge of these systems has grown rapidly in recent years, it is clear that a full understanding of cellular recognition phenomena will require an integrated and multidisciplinary approach.

This series aims to expedite such an understanding by bringing together accounts by leading researchers of all biochemical, cellular and evolutionary aspects of recognition systems. This series will contain volumes of two types. First, there will be volumes containing about five reviews from different areas of the general subject written at a level suitable for all biologically oriented scientists (Receptors and Recognition, series A). Secondly, there will be more specialized volumes (Receptors and Recognition, series B), each of which will be devoted to just one particularly important area.

Advisory Editorial Board

Receptors and Recognition

Receptors and
Recognition

Series B Volume 11

Membrane
Receptors

Methods for Purification
and Characterization

Edited by
S. Jacobs
and
P. Cuatrecasas

Wellcome Research Laboratories,
Research Triangle Park,
North Carolina, U.S.A.

LONDON AND NEW YORK
CHAPMAN AND HALL

First published 1981
by Chapman and Hall Ltd.,
11 New Fetter Lane, London EC4P 4EE

Published in the U.S.A. by
Chapman and Hall
in association with Methuen, Inc.
733 Third Avenue, New York, NY 10017

© *1981 Chapman and Hall*

Typeset by C. Josée Utteridge-Faivre
and printed in Great Britain
at the University Press, Cambridge

ISBN 0 412 21740 6

British Library Cataloguing in Publication Data

Receptors and recognition.
   Series B. Vol. 11: Membrane receptors.
   1. Cell interaction
   I. Jacobs, S.   II. Cuatrecasas, Pedro
   574.87'6   QH604.2   80–41218.

   ISBN 0-412-21740-6

Contents

Contributors

E. Bock, The Protein Laboratory, University of Copenhagen, Denmark.

O.J. Bjerrum, Department of Biochemistry and Molecular Biology, Northwestern University, Evanston, Illinois, U.S.A.

T.C. Bøg-Hansen, The Protein Laboratory, University of Copenhagen, Denmark.

H.R. Bourne, Division of Clinical Pharmacology, Departments of Medicine and Pharmacology, Cardiovascular Research Institute, University of California, San Francisco, U.S.A.

P. Coffino, Division of Clinical Pharmacology, Departments of Medicine and Microbiology/Immunology, University of California, San Francisco, U.S.A.

P. Cuatrecasas, Wellcome Research Laboratories, Research Triangle Park, North Carolina, U.S.A.

M. Das, Department of Biochemistry and Biophysics, University of Pennsylvania School of Medicine, Philadelphia, U.S.A.

E.L. Elson, Department of Chemistry, Cornell University, Ithaca, New York, U.S.A.

C.F. Fox, Department of Microbiology and the Parvin Cancer Research Laboratories, Molecular Biology Institute, University of California, Los Angeles, U.S.A.

G.M. Hebdon, Wellcome Research Laboratories, Research Triangle Park, North Carolina, U.S.A.

M.D. Hollenberg, Division of Pharmacology, University of Calgary, School of Medicine, Alberta, Canada.

S. Jacobs, Wellcome Research Laboratories, Research Triangle Park, North Carolina, U.S.A.

G.L. Johnson, Section of Physiological Chemistry, Division of Biology and Medicine, Brown University, Providence, Rhode Island, U.S.A.

H. Le Vine, Wellcome Research Laboratories, Research Triangle Park, North Carolina, U.S.A.

P.S. Linsley, Department of Microbiology and the Parvin Cancer Research Laboratories, Molecular Biology Institute, University of California, Los Angeles, U.S.A.

E. Nexø, Department of Laboratory Medicine, Johns Hopkins University, School of Medicine, Baltimore, Maryland, U.S.A.

J. Ramlau, The Protein Laboratory, University of Copenhagen, Denmark.

J.A. Reynolds, Whitehead Medical Research Institute and Department of Biochemistry, Duke University Medical Center, Durham, North Carolina, U.S.A.

N.E. Sahyoun, Wellcome Research Laboratories, Research Triangle Park, North Carolina, U.S.A.

J. Schlessinger, Department of Chemical Immunology, The Weizmann Institute of Science, Rehovot, Israel.

C. Schmitges, Wellcome Research Laboratories, Research Triangle Park, North Carolina, U.S.A.

Preface

Hardly a decade ago, membrane receptors were an attractive but largely unproven concept. Since that time enormous progress has been made, and we are now able to consider receptors much more concretely. Their existence has been established, their binding properties have been determined, and in some cases, they have been highly purified and their physical-chemical properties studied. It is now even possible to visualize microscopically some receptors. This progress has resulted largely from the development of highly powerful methods. These methods are the subject of this volume.

Although considerably diverse, different receptors share certain common properties, and common problems are encountered in their study. Consequently, a small number of techniques are particularly useful in studying different types of receptors. Thus, it makes sense to speak about membrane receptor methodology.

A very apparent problem in the study of membrane receptors is their presence in exceedingly small quantities and in a highly impure state. Therefore, very sensitive and specific techniques are required for their detection, characterization and purification. Such sensitivity and specificity is provided by the ability of receptors to bind certain ligands with very high affinity, and it is not surprising that most of the methods described in this volume depend upon this high affinity binding. The antigen-antibody interaction is of comparable sensitivity and specificity. Recently, a number of anti-receptor antibodies have been produced or found to occur spontaneously in auto-immune diseases. Undoubtedly, more will be produced in the future. The development of techniques to produce monoclonal antibodies *in vitro* is likely to enhance this trend. This should make the quantitative analytical immunologic methods described by Bjerrum *et al.* increasingly valuable.

Another aspect common to membrane receptors is that they are integral membrane proteins. This presents two problems. First, in order to fully understand their behavior it is necessary to determine their organization and dynamic arrangement within the membrane. Recent studies using the electron-microscopic and fluorescent techniques, described by Schlessinger and Elson, have clearly emphasized the importance of these factors. Second, in order to purify a receptor and to determine several of its physical chemical properties, it is necessary to extract it from the membrane. The chapter by Reynolds describes the scientific basis of the art of solubilizing receptors and characterizing them in the presence of detergents. Not only is it desirable to take receptors out of the membrane, but one of the ultimate goals of membrane receptor research is to reconstitute active receptor systems from isolated, purified components. Hebdon, *et al.* describe the tools currently available to accomplish this. Another technique which has been particularly useful in analyzing multicomponent receptor systems, and which shortcuts the need for physical separation of components and reconstituting them involves a genetic approach. Johnson *et al.* describe the use of this technique.

In this volume, no attempt has been made to provide exhaustively referenced review articles, nor have procedural details been presented in cook-book fashion in detail sufficient to allow the laboratory worker to actually carry out the procedures. Such detail is available in the cited literature, and will probably vary with the particular system studied. Rather, an attempt has been made to describe the underlying principles behind the various methods, to indicate in which situations they will be most useful, and to point out their potential limitations and drawbacks. We hope this will be of use to readers interested in understanding recent advances in the receptor field, as well as to researchers who wish to apply these methods to the particular receptor system that they are studying.

Wellcome Research Laboratories   S. Jacobs
North Carolina   P. Cuatrecasas

1 Receptor Binding Assays

MORLEY D. HOLLENBERG AND EBBA NEXØ

Membrane Receptors: *Methods for Purification and Characterization*
(*Receptors and Recognition*, Series B, Volume 11)
Edited by S. Jacobs and P. Cuatrecasas
Published in 1981 by Chapman and Hall, 11 New Fetter Lane, London EC4P 4EE
© Chapman and Hall

1.1 INTRODUCTION

The rapid advance in the understanding of the receptor mediated mechanisms, whereby a variety of drugs, hormones and toxins act at the cell membrane, can be attributed in large part to the development of reliable ligand binding assays. Although the methods employed in receptor related studies do not differ in principle from techniques used to measure ligand binding in other biochemical systems, the high affinity, low capacity and strict chemical specificity of pharmacologic receptors have presented a special challenge. Since the landmark studies of Clark (Clark, 1926a, b; 1933), it has been recognized that membrane receptors for drugs or other agents are present in vanishing small numbers (about 10^5 molecules per cell or fewer) compared to the amounts of other constituents in cell membranes. Thus, although in the past sensitive bioassay methods have yielded some limited quantitative data related to drug receptor interactions, precise measurements require the use of highly radioactive biologically active ligand probes. Further, it has been necessary in studies of ligand binding to distinguish 'specific' or receptor related binding from the 'non-specific' binding of radiolabeled ligand to structures other than the receptor. As summarized in a relatively recent review article (Cuatrecasas and Hollenberg, 1976), there are now available a variety of methods for the preparation of suitable ligand probes; further it has been possible to develop a number of criteria to aid in distinguishing between 'specific' and 'non-specific' binding in receptor studies (Hollenberg and Cuatrecasas, 1979). Thus, for the purposes of this chapter, it will be assumed that a suitable ligand probe may be available for a study of interest and that the distinction of receptor from non-receptor binding will not prove a problem. Attention will be focused on the methods that have proved successful in the study of membrane receptors, both in the particulate and soluble state. Since the methods outlined yield information not only about the receptor, but also about the ligand probe, consideration will also be given to the use of receptor preparations for radioreceptor assays.

1.2 GENERAL CONSIDERATIONS

The approach to the study of a receptor of interest can be relatively straightforward. Nonetheless, for each new ligand studied, a few preliminary considerations are often instructive. Firstly, it is important to know that the radioactive ligand probe chosen for study possesses a specific radioactivity sufficient to detect the expected number of receptor sites. As a rule, target cells possess from 10^4 to 10^5 binding sites per cell; in membrane preparations from receptor-rich organs it is not

3

unreasonable to find binding capacities in the range of 0.2 to 20 pmol mg^{-1} membrane protein. Thus, for a target cell having 10^4 sites per cell, a ligand possessing a specific activity of 10–20 Ci mmol^{-1} would yield a maximum binding of only a few hundred disintegrations per min for a sample size of 10^6 cells; there is thus considerable advantage in working with ^{125}I or ^{131}I-labeled derivatives, having specific activities in excess of 1000 Ci mmol^{-1}.

In choosing a method for study, it is also important to consider the rates of ligand binding. For instance, ligands like insulin, with equilibrium dissociation constants lower than 10^{-9} M exhibit dissociation rate constants (k_{-1}) of about 10^{-3} s^{-1}, corresponding to a half-life in excess of 10 min ($T_{1/2} = 0.693/k_{-1}$) for the ligand–receptor complex. In contrast, for compounds with receptor affinities lower than that of insulin (e.g. K_d 10^{-8} M), but with similar on-rate constants ($k_1 \simeq 10^7$ M^{-1}s^{-1}), a half-life of less than 7 s would be predicted for the receptor–ligand complex. Since the dissociation rate can be prolonged up to ten-fold simply by lowering the temperature, it might prove essential in such cases to 'freeze' the binding reaction at equilibrium by rapidly chilling the reaction mixture and performing all subsequent operations at 4°C. Alternatively, studies of ligands with comparatively low receptor affinities might necessitate the use of methods such as equilibrium dialysis or equilibrium gel filtration (Hummel and Dreyer, 1962).

A third consideration in the design of receptor binding assays relates to the integrity both of the receptor and the ligand. Firstly, for radioactively labeled ligands, it is essential to know the proportion of radioactivity that represents the intact active ligand. Ideally, 100% of the radioactivity should be present as the active ligand; however, with iodinated peptides, storage even at low temperatures (−70°C) over comparatively short time periods (2–3 weeks) may not prevent the formation of inactive iodinated peptide fragments. Despite this difficulty, with suitably high concentrations of either antibody or (preferably) receptor, it is possible to estimate the maximum amount of ligand that can be bound, so as to determine the amount of active labeled compound present in a preparation. For instance, with a radioactive ligand present at a concentration of 10^{-10} M, a receptor exhibiting a K_d of about 10^{-10} M should be capable of binding nearly all of the radioactivity ($\geqslant 99\%$), provided the receptor concentration can be raised to about 10^{-8} M. A good example of the use of receptor binding activity to estimate the proportion of active labeled ligand can be seen in studies with gonadotropic hormones (Catt *et al.*, 1976). In such studies, accurate estimates of receptor affinity can be obtained, despite the use of gonadotropin preparations exhibiting less than the maximum attainable biological activity. Failure to determine the proportion of active radioligand present will lead to an overestimate of the amount of 'free' hormone present and an underestimate of the 'bound' hormone concentration, so as to lead to serious errors in the estimates of ligand affinities by any chosen method of data analysis (e.g. see Chapter 2).

A second problem connected with the integrity of the ligand and its receptor concerns the degradation of both species that can occur during the course of binding studies. Because of potent proteases that are present in both intact cell and purified membrane—receptor preparations, binding studies performed at 37°C may greatly underestimate both the ligand affinity and the ligand binding capacity. Furthermore, storage of membrane preparations, even at 0°C, can lead to receptor degradation that is not blocked by conventional inhibitors of proteolytic enzymes. The extent of ligand degradation during the binding assay can be assessed either by estimating the ability of unbound ligand to bind to a second aliquot of membranes or by examining other physical characteristics of the unbound ligand (e.g. precipitability with trichloroacetic acid, gel filtration, electrophoretic or chromatographic mobility). The integrity of the receptor may be estimated by measuring the binding of receptor previously incubated in the absence of ligand under the binding assay conditions. Fortunately, in many instances, experiments done at about 24°C appear to minimize ligand and receptor degradation, and permit sufficiently rapid time courses of binding equilibrium.

A concern in membrane receptor studies that is encountered only infrequently in studies with enzymes relates to the absolute concentration of receptor present in a test system. Because of the high ligand affinity of many receptors ($K_d \simeq 10^{-9}$ M), it is possible, in certain instances, to achieve receptor concentrations equal to or even greater than the ligand equilibrium dissociation constant. In such instances, receptor concentration must be taken into account for design of experiments and for the analysis of binding data. For enzyme kinetics, an analogous situation is seen in the so-called 'zone behaviour' of enzymes in the presence of inhibitors (Straus and Goldstein, 1943).

Provided the ligand—receptor interaction obeys a simple mass action relationship (see, however, Chapter 2), the equilibrium dissociation constant, K_d, will be given by the equation:

$$K_d = \frac{[L] \times [R]}{[RL]},$$

wherein $[L]$, $[R]$, and $[RL]$ represent the concentrations of *free* ligand, the unoccupied receptor concentration and the ligand—receptor complex concentration respectively. Provided the concentration of *free* ligand and of ligand—receptor complex can be measured accurately, the K_d can be determined according to a number of rearrangements of the above equilibrium equation. For instance, a plot of *free* ligand concentration versus the concentration of bound ligand yields the K_d as the abscissa value for which binding is half-maximal. The K_d can be estimated directly from such binding isotherms.

Alternatively, as is frequently done, the data can be analysed according to the method of Scatchard (1949) (Fig. 1.1). Hypothetical binding curves based on the above equations are illustrated in Fig. 1.1, along with the corresponding Scatchard plots of the data. Although in principle the concentration of receptor should not affect the values obtained by this plot of the data, practically, the receptor concentration will have a bearing on experiments designed to obtain data from

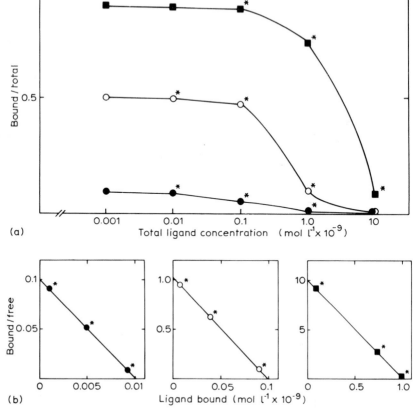

(a)

(b)

Fig. 1.1 Theoretical binding data and Scatchard Plots for studies with
receptor concentrations ranging from 1/10 to 10 times the K_d. (a) Fraction
of radioligand bound (ordinate) at ligand concentrations ranging from 0.001
to 10 nM (abscissa, logarithmic scale). A receptor K_d of 0.1 nM is assumed
and data are calculated from the mass action equation for receptor concentrations of : 1.0×10^{-9} mol^{-1} (■—■); 0.1×10^{-9} mol^{-1} (○—○);
0.01×10^{-9} mol^{-1} (●—●). (b) Data points indicated by asterisks (*) were
used to plot the data according to Scatchard (1949).

Scatchard plots. Ideally, to obtain the K_d for a ligand interaction, data should be
obtained over a range of total ligand concentrations sufficient to saturate between
about 10% and 90% of the receptor. As a rule of thumb this concentration range
corresponds to values from 1/10 to 10 times the magnitude of *either* the K_d *or* the
receptor concentration, depending on whichever value is the larger; the consequence
of increasing receptor concentration, in relation to the increased total ligand

concentration required to achieve receptor saturation is illustrated in Fig. 1.1a.
In Fig. 1.1b it can be seen that, provided a suitable range of *total* ligand concentrations is selected, analysis of the data according to Scatchard (1949) yields acceptable results.

The receptor concentration used will also affect the precision of the binding data obtained. For instance, at low receptor concentrations, the limiting factor is the accuracy with which the small amount of ligand bound can be measured. It can be calculated that for a receptor concentration 1/10 that of the K_d, ligand concentrations ranging from 1/10 to 10 times the value of the K_d will result in the binding of between 1% and 8% of the total ligand present (Fig. 1.1a, lower binding curve). The limitations of the expected binding data can thus be estimated from a knowledge of the specific activity of the radioligand and of the expected experimental background (i.e. non-specific binding).

At comparatively high receptor concentrations (e.g. 10 times the K_d value), virtually all of the ligand may be bound at low (relative to the K_d) ligand concentrations, such that the estimate of the amount of *free* ligand can prove difficult (Fig. 1.1a, upper binding curve). Nonetheless, when ligand concentrations ranging between 1 and 100 times the K_d are used, from 10% to 90% of the ligand will be bound and an accurate Scatchard plot can be obtained. In practice, receptor concentrations ten-fold higher than the K_d are only rarely achieved. In view of the above considerations, it can be suggested that optimal binding data may be obtained for receptor concentrations equivalent to the K_d and for ligand concentrations ranging from 1/10 to 10 times the K_d (Fig. 1.1a, middle binding curve). The amount of ligand bound will range from 10% to 50% of the total, so as to permit accurate measurements without the use of excessive concentrations of either the ligand or the receptor.

In binding studies wherein the affinities of one or more analogues of the radiolabeled compound are examined, affinities are frequently calculated from the concentration of unlabeled ligand at which the binding of radiolabeled ligand is reduced to 50% of maximum (the so-called IC_{50}). It is readily apparent that the concentration and receptor affinity of the labeled ligand must be known in order to calculate the K_d from the IC_{50} as is often done according to the equation (Cheng and Prussoff, 1973):

$$K_d = IC_{50} / (1 + \frac{L^*}{K_d^*})$$

where L^* and K_d^* represent respectively the concentration and dissociation constant of labeled ligand. It may not be appreciated that the above formula is valid *only* if the receptor concentration is low relative to the K_d. The importance of receptor concentration in such studies has been taken into account by Jacobs *et al.* (1975).

The free unlabeled-ligand concentration at which 50% of the bound labeled ligand will be displaced (IC_{50}) by unlabeled ligand can be defined as K_d (app).

If the labeled and unlabeled ligand possess identical receptor affinities, it can be shown (Jacobs *et al.,* 1975) that:

$$K_d \text{ (app)} = K_d + L^* + R_t - \frac{3}{2}RL^*$$

wherein K_d represents the true ligand dissociation constant, L^* is the total concentration of radioligand, R_t is the net receptor concentration and RL^* represents the concentration of radioligand bound in the *absence* of unlabeled ligand. Thus, only when $K_d \gg (R_t + L^*)$ does the K_d (app) (or IC_{50}) closely approximate the dissociation constant of the unlabeled ligand.

Receptor concentrations that approach the value of the K_d can lead to anomalous results when the kinetics of binding are studied. When the ratio of binding site concentration to the dissociation constant ($[R]/K_d$) exceeds a value of 10, a 'retention effect' may be observed, wherein the rate of efflux of a radioactive ligand from a biological structure (membrane, cell surface, nucleus) is slower in the absence than in the presence of an excess of unlabeled ligand. The 'retention effect' has been described by Silhavy *et al.* (1975) in connection with studies of the binding of maltose to the periplasmic maltose binding protein of *E. coli.* In studies with membrane receptors this effect, amongst other possibilities (e.g. negative co-operativity of binding), must be considered where the spontaneous radioligand off-rate differs when measured either in the presence or absence of an excess of unlabeled ligand.

The essence of the above discussion is that for each ligand–receptor combination of interest, it is necessary to select a sufficiently radioactive, biologically-active ligand probe, to optimize the receptor concentration in keeping with the expected ligand affinity, to study an appropriate range of ligand concentrations, to select a temperature that minimizes both receptor and ligand degradation but yet permits convenient binding equilibration times, and to choose a method of separating free from receptor-bound ligand that is appropriate for the expected ligand off-rate. These preliminary considerations are of as much importance as the concerns related to the binding criteria (appropriate ligand affinity, saturability, reversibility, concordance of chemical binding specificity with bioactivity, and tissue specificity) that, as discussed elsewhere (Cuatrecasas and Hollenberg, 1976; Hollenberg and Cuatrecasas, 1979), are expected of a receptor-related binding interaction.

1.3 METHODS FOR RECEPTOR DETECTION AND QUANTITATION

In ligand binding measurements with either particulate or soluble membrane preparations, it is necessary to determine the proportion of free and receptor-bound ligand. With intact cells or membranes, a rapid and efficient separation of membrane-bound ligand from the supernatant medium is possible. Since, as discussed above, the binding capacity of membrane or cell preparations is very low and, since in many binding assays less than 10% of the total ligand is bound, the most precise data are obtained by measuring directly the amount of membrane-bound ligand, rather than by obtaining such data from the difference between the amount of free ligand present before and after the addition of receptor. The separation of particulate material from the aqueous phase must be achieved rapidly and in a manner that provides for thorough washing, without undue trapping of radioactivity from the aqueous phase and without appreciable non-specific adsorption of the ligand to the apparatus used (e.g. plastic or glass tubes, glass or synthetic polymer filters). As discussed above, the rapidity with which separation of free from bound ligand must be achieved can be estimated by approximating the half-life of the ligand receptor complex on the basis of an expected dissociation constant.

1.3.1 Studies with intact cell or membrane preparations

(a) *Centrifugation*
Perhaps the simplest and most economical technique for measuring the binding of ligands to cells employs centrifugation to separate bound from free ligand. The binding reaction is allowed to proceed to equilibrium (usually for 30–40 min at 24°C) and aliquots of the cell or membrane suspension are rapidly chilled to 4°C and pelleted. The resulting pellet is quickly rinsed with cold buffer without resuspension and radioactivity in the particulate material is measured either directly (e.g. ^{125}I) or after solubilization (^{14}C, ^3H). Many variations of this technique are possible, especially with high-affinity systems where the half-life of the ligand – receptor complex is prolonged at 4°C. Provided the trapping of fluid in the pellet is minimal and identical for all samples, the difference in radioactivity bound in the presence and in the absence of a large excess of unlabeled ligand will reflect the 'specific' ligand binding. For convenience, the entire procedure may be performed in 12 x 75 mm disposable glass test tubes; alternatively, aliquots of the reaction mixture may be transferred to plastic 'microfuge' tubes (either 0.4 ml or 1.5 ml capacity tubes are available) for harvesting and rapidly rinsing (with buffer at 4°C) the membrane or cell pellet. When plastic tubes are used, the tips containing the compact pellet can be cut from the remainder of the tube for the measurement of radioactivity. It should not be overlooked that many radioactive ligand derivatives may adsorb strongly and in a displaceable manner to plastic tubes, so as to mimic receptor binding in the absence of membrane, thereby complicating measurements by this technique. Such a complication should be ruled out in the initial stages of any study.

The centrifugation technique can be refined with the use of mixtures of miscible oils of different density (Gliemann *et al.*, 1972; Chang and Cuatrecasas, 1974; Livingston *et al.*, 1974). Dinonyl phthalate (ICN Laboratories), which possesses a specific gravity less than that of water but greater than that of fat cells, can be used to separate fat cells from an aqueous suspension by flotation above the water/oil interface (Gliemann *et al.,* 1972). Fat cell suspensions are transferred to small conical polyethylene tubes containing about 200 μl of oil and centrifuged briefly (30 s) in a small, high-velocity, rapidly-accelerating centrifuge (e.g. Beckman 152 or Eppendorf microcentrifuge). The tube is then cut between the fat cell layer and the oil/water interface (through the oil phase) and the amount of bound radioactivity is measured. Dibutyl phyhalate, which is heavier than water, can be mixed with dinonyl phthalate in proportions (e.g., dibutyl : dinonyl phthalate, 2 : 1 v/v) which yield a mixture heavier than water but lighter than many cell or membrane preparations. In a manner similar to that described above, many cell and membrane preparations can be quickly spun down below the water/oil interface, free from the aqueous medium, so as to measure the mebrane-bound radioactivity.

In practice, the binding reaction is frequently performed in disposable glass tubes; at binding equilibrium, 0.2 ml aliquots of the membrane suspension are pipetted into 0.4 ml plastic microcentrifuge tubes that have been pre-filled with 0.1 ml of the 2:1 v/v oil mixture described above. After 1–2 min of centrifugation (e.g. at approximately 10 000 g at 24°C in a Beckman 'Microfuge B'), the cells or membranes can be seen as a compact pellet below the water/oil interface. The tube tips are cut through below the oil/water interface in order to measure the radioactivity in the pellet. The supernatant solution can also be recovered for further analysis if necessary.

The oil/water centrifugation technique separates free from bound ligand extremely rapidly, and is the method of choice for the study of binding to 'low affinity' sites with a short half-life of dissociation. Since, during centrifugation, the cells or membranes are in continuous contact with the supernatant fluid, the rapidity with which a separation (between particulate and fluid phase) is achieved is governed essentially by the speed with which the pellet is rinsed by ligand-free buffer (possibly several seconds) or by the speed with which the particulate material crosses the oil/water interface (certainly less than 1 s).

It is often a problem when ^{3}H- or ^{14}C-labeled ligands are used, to dissolve the compact cell pellet completely. It is advantageous to use a scintillation medium containing a solubilizer (e.g. Beckman BBS-3) or to measure radioactivity after combustion. As a minimal precaution, the pellet should be left overnight in the scintillation medium to allow for disruption of the pellet.

(b) *Filtration*

Filtration of particulate material on synthetic polymer or porous glass filters provides another rapid, efficient way to separate free from membrane-bound ligand. The choice of an appropriate filter is critical, since many compounds, such as glucagon and insulin (Cuatrecasas and Hollenberg, 1975; Cuatrecasas *et al.,* 1975) can, at very low concentrations, adsorb in a displaceable manner to materials such as glass and cellulose acetate. Each ligand necessitates a search for an appropriate filter [e.g. for insulin and epidermal growth factor–urogastrone (EGF–URO), Millipore series E (cellulose acetate); for wheat germ agglutinin, series N (nylon);

for Concanavalin A, series L (Teflon)] yielding the lowest 'background' non-displaceable binding; equivalent filters from Amicon Corporation have also proved effective for studies with EGF—URO. Filters are immobilized in a multiple manifold apparatus with a tight-fitting silicone rubber washer sealing the membrane against its support. The seal must be free of dead space around the periphery so that washing will be complete. A suspension of cells or membranes (0.2–0.5 ml) is filtered (1 μm pore size of filters for cells, 0.2 μm for membranes) under reduced pressure, and the collected material is rapidly (10–15 s) washed with 10 ml of ice-cold buffer containing 0.1% w/v bovine serum albumin. The albumin, present both in the incubation medium and washing the buffer, minimizes the non-specific adsorption of the ligand to filters and vessel walls. The radioactivity trapped on the filter is then measured; if desired, the filtrate can also be collected for analysis. When ^{3}H-labeled ligands are studied, it is particularly important to solubilize the membranes and filter completely. Filters can be shaken overnight in a small volume (1 ml) of 10% w/v aqueous sodium dodecyl sulfate (SDS) before addition of a high-efficiency scintillant for measurement of radioactivity by scintillation counting. Measurement of radioactivity after combustion of the sample provides an alternative.

One disadvantage of the porous synthetic-polymer filters is that large amounts of cells or membranes can markedly reduce the filtration rate. The filtration of membrane preparations can be facilitated by adding polyethylene glycol (Carbowax 6000, final concentration 10% w/v) immediately prior to filtration, so as to aggregate the membrane components. Alternatively, the use of glass filters (e.g. Whatman series GF—A or GF—B) can circumvent this problem, provided non-specific adsorption of the ligand to such filters does not interfere. Such filters have been used with great success in the study of binding of opiate analogues to membranes from the central nervous system (Snyder *et al.*, 1975).

(c) *Gel filtration*
Although gel filtration is frequently used in connection with studies of soluble receptors, this method may also prove of value in the study of membrane vesicles. For instance, short columns of Sephadex G—50 permit erythrocyte vesicles prepared by sonication to pass through in the void volume. Thus, it was possible in studies of adrenergic receptors in toad erythrocyte vesicles to separate free from membrane-bound ligand (^{125}I-hydroxybenzylpindolol) by gel filtration of 0.1 ml aliquots on 0.5 x 8 cm columns of Sephadex G—50 at 4°C (Sahyoun *et al.*, 1977). At 4°C, the ligand dissociation rate is slow enough (at 30 min less than 20% of the label dissociates) so that essentially all of the vesicle-bound ligand appeared in the column void volume). Presumably this method of assay could work for other membrane ligand systems with sufficiently prolonged ligand dissociation rates.

(d) *Equilibrium dialysis*
Equilibrium dialysis can also be used to study ligand membrane interactions. Small dialysis cells (e.g. 0.3 ml) separated by an appropriate membrane provide for rapid

equilibration (about 3 h) of a small ligand such as isoproterenol (Atlas *et al.,* 1974; Levitzki *et al.,* 1974, 1975). This technique is particularly suited to ligands with relatively low receptor affinities (K_d = 10^{-7}M) for which the very short half-life of the receptor–ligand complex would preclude measurements by techniques previously discussed. The binding of (^{3}H-labeled)-isoproterenol and (^{3}H-labeled)-propranolol to turkey erythrocyte ghosts, measured by equilibrium dialysis, reveals a saturable component, with half-maximal saturation at 0.15 μM isoproterenol (Atlas *et al.,* 1974; Levitzki *et al.,* 1974, 1975). The saturable component comprises only a small proportion of the overall binding isotherm and, as in other studies, the specific catecholamine binding was obtained by subtracting from the total bound that which remained in the presence of an excess of the competitive antagonist, propranolol. The precision of such studies is limited not only because of the large proportion of radioactivity that remains bound in the presence of unlabeled competing ligand but also because the data are derived by subtraction of two relatively large measured values to yield the (small) amount of bound ligand.

1.3.2 Soluble receptors

In many cases, it has been possible with the use of detergents such as Triton X-100 to solubilize from membrane preparations receptors that retain their ability to bind ligands with specificity and high affinity. The measurements of such receptor interactions present special technical problems, in view of the considerations outlined above for the studies with particulate receptors. It is more difficult to achieve a rapid separation of free from soluble receptor-bound ligand. Furthermore, the presence of detergent at concentrations sufficient to keep the receptor in solution may interfere with ligand binding. In general, advantage is taken of the substantial physicochemical differences (charge, solubility, molecular weight, etc.) between the free (usually low molecular weight) ligand and the ligand–receptor complex. The presence of detergent introduces the complication that the detergent micelles of high molecular weight can bind free ligand, so as to make separation from receptor-bound ligand difficult. Fortunately, the apparent molecular weight of the Triton micelle (about 90 000) is smaller than that of receptors so far studied in Triton-containing solutions.

(a) *Polyethylene glycol precipitation*
Since its introduction as a differential precipitating agent for fractionating proteins (Polson and Deeks, 1963; Polson *et al.,* 1964; Chesebro and Svehag, 1968), polyethylene glycol has been useful in isolating large macromolecular species, including viruses (Polson and Deeks, 1963; Herbert, 1963). In general the solubility of protein decreases with increasing molecular weight (Juckes, 1971). In particular, polyethylene glycol (Carbowax 6000, Union Carbide) has proved to be useful to separate free and antibody-bound peptide hormones in radioimmunoassays (Desbuquois and Aurbach, 1971), and a similar method has been developed to

measure insulin binding to soluble receptors (Cuatrecasas, 1972a,b,c). It is expected that for each hormone studied, different conditions will be necessary to precipitate the maximum amount of hormone−receptor complex, leaving most of the free ligand in solution. The procedure that has been used for the insulin receptor may serve as a prototype for other studies.

For measurement of insulin binding, 5−50 μl of detergent-extracted supernatant is added to 0.2 ml. Krebs-Ringer bicarbonate buffer−0.1% w/v albumin, pH 7.4 containing (^{125}I) insulin with (control tubes only) or without 25−50 μg of native insulin per milliliter. Phosphate buffer (0.1 M, pH 7.4) may also be used for the binding assay provided the pH is maintained between pH 7.0−7.4; buffers containing Tris−HCl, however, appear to interfere with the polyethylene glycol precipitation. Equilibration of binding is achieved in about 20−50 min at 24°C, at which time 0.5 ml of ice-cold 0.1 M sodium phosphate buffer (pH 7.4) containing 0.1% bovine γ-globulin is added and the tubes are placed in ice. Cold 25% w/v polyethylene glycol (0.5 ml) is added (final concentration, 10% w/v), and the tubes are thoroughly mixed and placed in ice for 10−15 min. The suspension is then filtered under reduced pressure on cellulose acetate (Millipore EG or EH) filters, and the collected precipitate is washed with 3 ml of 8% w/v polyethylene glycol in 0.1 M phosphate buffer (pH 7.4) before measurement of trapped radioactivity by crystal scintillation counting. As for the measurments of binding with cells and membranes, the specific binding is determined by subtracting from the total radioactivity bound, that which remains bound in the presence of a high concentration (25−50 μg ml^{-1}) of native insulin. Under the above conditions, less than 0.5% of the total free (^{125}I) insulin is precipitated or adsorbed non specifically and nearly quantitative precipitation of the insulin−receptor complex occurs (Cuatrecasas, 1972b).

Concentrations of polyethylene glycol less than 8% (w/v) incompletely precipitate the complex; concentrations higher than 12% significantly precipitate free insulin. The presence of γ-globulin is essential as a carrier for the precipitation reaction but concentrations above 0.1% (w/v) cause precipitation of free insulin. If the pH of the buffer containing the γ-globulin is above 8 or below 7, the complex is less effectively precipitated; phosphate buffers (0.1 M, pH 7.4) can be used effectively in the incubation medium. A final concentration of Triton X-100 in the assay mixture in excess of 0.2% v/v results in decreased insulin binding. The membrane extracts are therefore diluted before assay so that the final concentration of Triton X-100 is usually less than 0.1% and always less than 0.2% (v/v). For the estimation of rate data (association/dissociation) it is assumed that the addition of the cold polyethylene glycol-γ-globulin solutions stop the binding reaction at the timed intervals.

The ability to precipitate the hormone−receptor complex and wash the precipitate with ligand-free buffer without appreciably losing the bound hormone may be attributed either to the reduction of the off-rate by the relatively low temperature of the wash procedure, and/or to the so-called 'retention effect' referenced above (Silhavy *et al.,* 1975), which occurs under conditions where the ratio of receptor concentration to the dissociation constant is large, e.g. $R/K_d \geqslant 10$. In principle, any analogous method for selective precipitation of the hormone receptor complex (ammonium sulfate or other salting-out agent; trichloracetic acid; change in pH) should provide a means of assaying soluble receptor, provided the complex is not dissociated by the precipitating agent. For example, salt fractionation has been used to examine the properties of soluble cholinergic receptor proteins (Franklin and Potter, 1972; Meunier *et al.*, 1972a,b). The analysis of soluble receptors for other ligands by precipitation technique should similarly be possible. It is apparent that if the half-life of the receptor–ligand complex is not sufficiently prolonged at low temperatures ($4°C$), the dilution of the sample (about five-fold in the procedure described above) and the time allowed for precipitation may shift the binding equilibrium appreciably; appropriate corrections of the data will then be necessary.

(b) *Equilibrium dialysis*

As indicated above, equilibrium dialysis provides an attractive technique for studying hormone–receptor interactions of relatively low affinity ($K_d \geqslant 10^{-7}$ M). This technique has been used successfully to measure binding of radioactively labeled cholinergic agents (acetylcholine, decamethonium, dimethyltubocurarine) to solubilized receptor proteins from the electric tissue of *Electrophorus electricus* (Meunier *et al.*, 1972a,b, 1974; Klett *et al.*, 1973, 1974) and *Torpedo* (Eldefrawi and Eldefrawi, 1973; Moody *et al.*, 1973). It should be noted, however, that receptor concentrations approaching the micromolar range were used for such studies, so as to generate appreciable differences in ligand concentration between the inside and the outside of the dialysis sac; such large amounts of soluble receptors for polypeptide hormones have not as yet been obtained in a pure state. It is assumed, moreover, in such studies that the binding of ligand to the dialysis membrane does not appreciably affect the measurements; this assumption is usually checked by the observation of ligand binding in the presence of a large amount of unlabeled antagonist (e.g. α-bungarotoxin, in excess of another radioactively-labeled cholinergic agent). The inherent error resulting from the subtraction of two larger values to yield the small amount of ligand bound is always present in such studies.

(c) *Gel filtration*

The separation of free from bound ligand in a solubilized receptor preparation can also be achieved by gel filtration. For example, chromatography of the soluble human placenta acceptor for human transcobalamin-II (HTCII) indicates that the TCII–^{57}Co Cobalamin complex can associate with the protein eluted just after the column void volume (Fig. 1.2; Nexø and Hollenberg, 1980); in the presence of an excess of unlabeled HTCII–Cobalamin, there is a reduction in radioactivity eluted in the high molecular weight fraction and a corresponding appearance of radioactivity at the elution volume of free HTCII. It is important to note that even at $4°C$, there can be spontaneous dissociation of the high molecular weight complex. For example, extensive dissociation might occur for ligand–receptor complexes possessing dissociation constants lower than that of the transcobalamin acceptor (K_d approximately 5×10^{-10} M). Similar data have been obtained for the binding of insulin to its receptor (Cuatrecasas, 1972b) and for the binding of radiolabeled neurotoxins to soluble nicotinic receptors for acetylcholine. The spontaneous dissociation upon chromatography even for these higher affinity receptors at 4°C indicates that precise quantitative binding data are difficult to obtain by this technique. Furthermore, the presence of detergent micelles (mol. wt. approximately 90 000; Kushner and Hubbard, 1954) to which ligand can adsorb may complicate gel filtration studies of receptor binding. Adsorption to a detergent micelle may have been responsible for complications in the study of soluble receptors for prolactin (Shiu and Friesen, 1976) which in detergent solutions migrated as a species with a molecular weight of about 80 000. It is therefore essential to demonstrate that the radioactivity in the high molecular weight fraction can be 'displaced' by an appropriately low concentration of unlabeled ligand. Despite the aforementioned difficulties, should the ligand affinity be high enough (and consequently, dissociation slow enough, e.g. at 4°C) rapid gel exclusion chromatography may yield acceptable quantitative results. For example, in work with soluble estrogen receptors, rapid chromatography at 4°C on columns of coarse Sephadex G−25 or G−50, yields excellent results (Sica *et al.,* 1973a,b).

Gel filtration can also be employed to obtain binding data with the use of the technique originally developed by Hummel and Dreyer (1962) and elaborated upon for quantitative analysis by Cuatrecasas *et al.* (1967). Briefly, the gel and eluting buffer are saturated with a low concentration of the radioactive ligand and the soluble macromolecule is then added to the column. Labeled ligand bound to the receptor is accumulated as a 'peak' of radioactivity, followed by an equivalent 'trough'. The amount of ligand bound can be calculated by measuring the amount of radioactivity in the 'peak' and 'trough'. This technique has been valuable for the measurement of glucagon binding to soluble receptor (Giorgio *et al.,* 1974). Small plastic columns (Kontes, Chromaflex) and Sephadex G−50, coarse; about 3.3 ml bed volume) are pre-equilibrated with buffer containing 0.1% bovine albumin and about 0.5 ng of ^{125}I-labeled glucagon per milliliter (approximately 50 000 ct min^{-1} ml^{-1}). The soluble glucagon binding protein ($2-50$ μl) is pre-equilibrated with 0.5 ml of the same buffer; the mixture is then applied to the gel and developed with about 5 ml of buffer containing (^{125}I-labeled) glucagon. The method provides a rapid microassay for monitoring the glucagon binding capacity of chromatographic fractions of solubilized liver membrane proteins. A similar technique has been used to measure the binding of ^{3}H-labeled tetrodotoxin to a soluble membrane extract of olfactory nerve (Henderson and Wang, 1972). In principle, this method may also prove applicable to the study of particulate receptor preparations, provided the membrane vesicles pass freely through the columns.

(d) *Lectin−agarose-immobilization assay*
It is the case that a large number of membrane receptors are glycoproteins which can

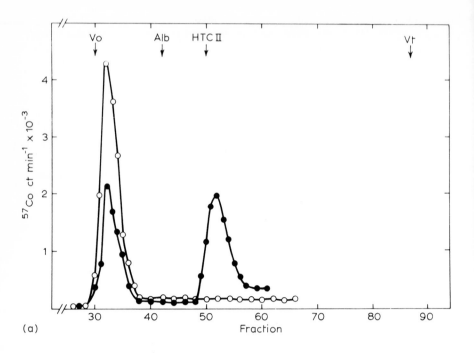

(a)

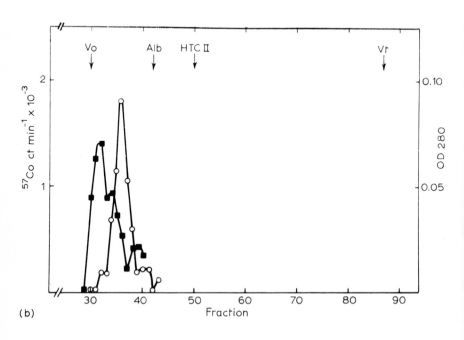

(b)

interact with sugar-specific plant lectins. In the course of our studies on the receptor for epidermal growth factor—urogastrone (EGF—URO) (Hock *et al.*, 1979a,b) and on the acceptor for transcobalamin-II (Nexø and Hollenberg, 1980), we were able to determine that subsequent to solubilization, both binding proteins could be adsorbed onto lectin—agarose derivatives and that in the lectin-immobilized state both binding proteins were still capable of specific ligand recognition. These observations led us to develop a new assay for soluble receptors (Nexø *et al.*, 1979) that has facilitated the characterization of the transcobalamin-II acceptor (Nexø and Hollenberg, 1980) and has led to the characterization and further purification of the receptor for EGF—URO (Hock *et al.*, 1979b).

The usefulness of the assay for the EGF—URO receptor is of particular interest, since it is not possible to detect ligand binding by the receptor in the soluble state using any of the methods discussed above. Apparently in the soluble state, the EGF—URO receptor loses its ligand recognition property. Upon immobilization of the solubilized receptor on concanavalin A—agarose, the ligand binding ability can again be detected. Since the method may prove useful specifically for the study of other receptors that apparently lose their ligand recognition property in the soluble state and may be of general use for the study of a variety of membrane proteins, the assay, outlined in Fig. 1.3, is described in some detail.

We have observed that the most readily available and useful lectin for the assay is concanavalin A (Con A). However, other lectins (e.g. wheat germ agglutinin) appear to work equally well and in principle it should be possible to optimize any assay of interest by an appropriate choice of lectin. Con A—Sepharose 4BCL (approximately 2 mg Con A per ml), prepared by the method of March *et al.* (1974), is used in 50 μl aliquots for the adsorption of replicate samples of soluble membrane

Fig. 1.2 Gel filtration of solubilized TCII—cobalamin acceptor from human placenta membranes. Solubilized acceptor from placenta membranes was first partially purified by adsorption to wheat germ agglutinin—agarose followed by elution with N-acetylglucosamine (0.2 mol l^{-1}). (a) Acceptor still adsorbed to the lectin column was incubated with TCII—^{57}Co-cobalamin for 1 h at 4°C prior to elution with the specific sugar. The eluate containing the TCII—^{57}Co-cobalamin acceptor complex was analysed either directly (○) or after incubation for 1 h at room temperature with 100 pmol pure un-labeled rabbit TCII (●) on columns (1.5 x 87 cm) of Sephacryl S—200 in 50 mM Tris—HCl, pH 7.4 containing 150 mmol l^{-1} sodium chloride, 1 mmol l^{-1} calcium chloride, 1 mmol l^{-1} manganese chloride and 0.1% w/v Ammonyx-LO, at a flow rate of about 7 ml h^{-1}; 1.67 ml fractions were collected. (b) Soluble acceptor (about 900 μg protein) was analysed by gel filtration, as above, and the TCII-cobalamin binding activity of the effluent fractions (○) was determined using Con A—Sepharose as outlined in the text. The optical density at 280 nm (●) of effluent fractions was also monitored. Column calibration: V_0, void volume; V_t, total volume; HTCII, human TCII-cobalamin complex; Alb, bovine serum albumin. (Data from Nexø and Hollenberg, 1980).

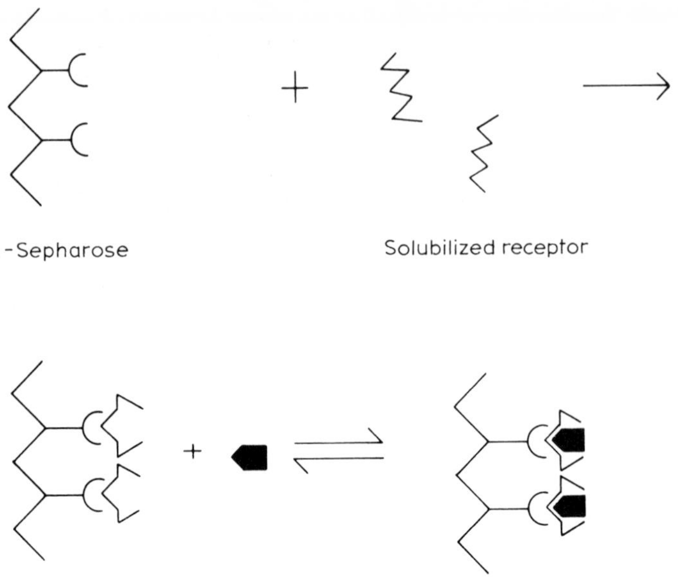

Fig. 1.3 Schematic representation of the lectin immobilization assay for solubilized receptors. A soluble receptor that cannot bind ligand in detergent solutions is immobilized on Con A–Sepharose. The immobilized receptor, in a suitable buffer, is again capable of binding the specific ligand.

protein (approximately 100 μg protein) in 12 x 75 mm glass tubes (15 min at 24°C; final reaction volume, approximately 300 μl). A variety of detergents (Triton X-100; Ammonyx–LO; Lubrol) do not interfere significantly with the adsorption of protein. The bead-bound protein is then washed by dilution with 5 ml ice-cold buffer, immediately harvested by centrifugation (2 min at 1000 rev min^{-1}), and is then incubated with radiolabeled ligand either with or without an excess (usually 10–100-fold) of unlabeled ligand. After equilibration (usually 1 h at 24°C), the samples are again washed by suspension in 5 ml ice-cold buffer and harvested by centrifugation (2 min at 2000 rev min^{-1}) for measurement of bead-bound radio-activity upon removal of the supernatant solution by aspiration. As in other studies, the 'specific' ligand binding is calculated by subtracting from the total radioactivity bound in the absence of unlabeled ligand, the amount of radioactivity bound in the presence of about a 100-fold excess of unlabeled ligand.

The method offers the advantages that the receptor can conveniently be trans-ferred *via* the beads to a buffer optimized for the measurement of ligand binding and that dilute receptor samples (e.g. column effluent fractions) can be concentrated on the beads for the study of ligand binding. An application of the method is illustrated in Fig. 1.2, wherein column effluent fractions have been assayed for the presence of the TCII acceptor. The method was of critical importance in this study, since neither the polyethylene glycol method nor ammonium sulfate precipitation could be used for the assay of the soluble TCII acceptor (TCII co-precipitates with the acceptor). This method should prove of use for the study of any receptor that recognizes a ligand free of carbohydrate and deserves particular consideration for the study of receptors that appear to lose their ligand recognition property in the soluble state.

(e) *Ion exchange*

Should the net charge of the hormone—receptor complex differ appreciably from that of the free-ligand, ion-exchange techniques can yield efficient separation of free from bound ligand. Based on this principle, a simple rapid procedure was developed for measuring the binding of strongly-cationic radioactively-labeled neurotoxins to soluble acetylcholine receptors (Klett *et al.*, 1973; Schmidt and Raftery, 1973a,b). The toxin—receptor complex is selectively adsorbed to DEAE—cellulose disks and the free ligand is washed off. As with other studies, it is essential to ensure that non-specific (non-electrostatic) binding of the ligand examined does not complicate the results. An ion exchange method has also proved useful in the study of the transcobalamin-II acceptor (Seligman and Allen, 1978). At 90 mM NaCl, 70% of the TCII—cobalamin acceptor could be bound to DEAE—cellulose, whereas only 10% of the free TCII was bound. The NaCl concentration of the buffer was, however, critical since a 10 mM decrease caused a three-fold increase in the adsorption of free TCII and a 10 mM increase caused a two-fold decrease in adsorption of the TCII—acceptor complex. Thus, the lectin immobilization assay described above provides a more versatile method for the study of the binding properties of the soluble TCII—acceptor.

(f) *Gel electrophoresis*

Although sodium dodecyl sulfate (SDS) appears to abolish the ligand recognition property of membrane receptors, polyacrylamide gel electrophoresis in Triton-containing buffers allows for the separation and detection of soluble membrane receptors. Provided the ligand affinity is sufficiently high and the ligand off-rate sufficiently slow, the radioactive ligand—receptor complex may be detected subsequent to electrophoretic analysis, either by autoradiography or by measuring the radioactivity in serial gel slices. As in other studies, it is important to demonstrate that unlabeled ligand can compete for the binding of labeled ligand to the electro-phoretic species thought to represent the receptor. Alternatively, the binding of ligand can be measured (e.g. by the polyethylene glycol assay described above) in fractions eluted from serial gel slices. Using gel electrophoresis, it has been possible to characterize a gonadotropin receptor (Catt *et al.*, 1976) and to detect two electrophoretic forms of the insulin receptor (Krupp and Livingston, 1978, 1979).

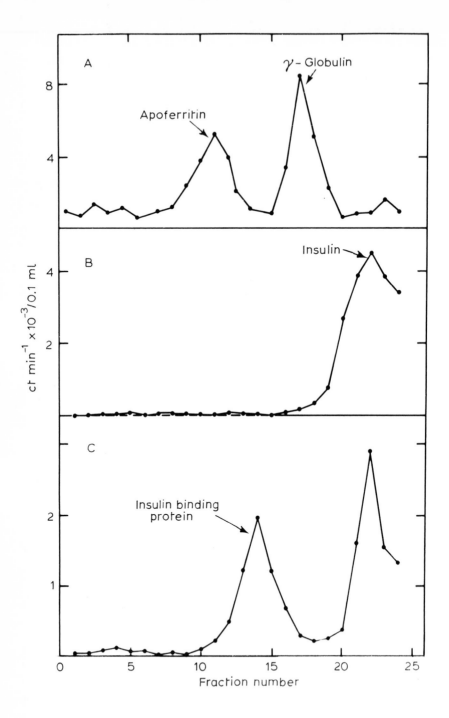

(g) *Sucrose gradient sedimentation*

Sedimentation in Triton-containing sucrose gradients provides another means of characterizing soluble membrane receptors. This method, which has been used extensively in the study of cytoplasmic receptors for steroid hormones is equally useful for the study of receptors for ligands such as insulin (Cuatrecasas, 1972a) and the catecholamines (Haga *et al.*, 1977). In brief, replicate samples of the soluble receptor preparation are equilibrated with radiolabeled ligand both with and without an excess of unlabeled ligand. The solutions are then layered on a sucrose gradient and the sedimentation velocity of receptor-bound radioactivity is determined. As indicated in the example with the insulin receptor (Fig. 1.4), it is necessary to show that the presence of an excess of unlabeled ligand eliminates the radioactivity sedimenting with the macromolecular species. As with the electrophoretic method described above, it is also possible to assay gradient fractions for ligand binding subsequent to centrifugation.

(h) *Fluorescence*

Most of the techniques for measuring hormone—receptor interactions employ radioactive derivatives for analysis. An alternative approach for the study of soluble and membrane-bound receptors has used fluorescent hormone derivatives as probes. Fluorescent derivative have been used extensively to study biological membranes and model systems (Waggoner and Stryer, 1970; Radda and Van Der Kooi, 1972). In particular, fluorescent derivatives have been used to study the interaction of cholinergic agents with cholinesterases (Mayer and Himmel, 1972; Mooser *et al.*, 1972) and with an acetylcholine-binding proteolipid (DeRobertis, 1971; Weber *et al.*, 1971); these derivatives have also proved useful for studies of the acetylcholine receptors. The fluorescence of one derivative, 1—(5-dimethylaminonaphthalene-l-sulfamido)-3-(trimethyl ammonium) propane iodide, is enhanced on interacting with acetylcholine receptor (Cohen and Changeux, 1973); the fluorescence of

Fig. 1.4 Sedimentation behavior of insulin binding protein of fat cell membranes on 5—20% sucrose gradients. Fat cell membranes were extracted with 1% (v/v) Triton X-100 and centrifuged for 70 min at about 100 000 g. The supernatant was dialysed for 16 h at $4°C$ against 0.1 M sodium phosphate buffer containing 0.2% (v/v) Triton X-100. The supernatant (0.2 ml, containing 0.5 mg of protein) was incubated at $24°C$ for 20 min with ^{125}I-insulin (3.5 x 10^4 ct min^{-1}) before being subjected to gradient centrifugation for 16 h. (c) Another sample of supernatant was processed identically except that native insulin (20 μg) was added 5 min before incubation with ^{125}I-insulin (b). A sample (0.2 ml) of the dialysis buffer described above was also centrifuged (a) under identical conditions after addition of ^{14}C-acetylapoferritin (6.1 x 10^4 ct min^{-1}) and ^{14}C-acetyl-γ-globulin (7.2 x 10^4 ct min^{-1}). The broader appearance of the insulin peak in (b) compared to that in (c) probably results from aggregated forms of insulin, which form in the former because of its high concentration. From Cuatrecasas (1972a).

another derivative, bis (3-amino-pyridine)-1, 10-decane, is quenched upon interacting with receptor. Measurements can be made, on the one hand, of the titration of receptor sites with the fluorescent probe and, on the other, of the competition by other cholinergic ligands for the binding of the fluorescent probe. The analysis of these interactions can be done in the same way as for radioactive probes (e.g. see Chapter 2) so as to yield dissociation constants for the fluorescent and native compounds. The data obtained by fluorescence methods compare favorably with those obtained using radioactive ligand derivatives (Cohen and Changeux, 1973; Martinez-Carrion and Raftery, 1973). The overall limitation of this technique stems from the quantitative considerations developed earlier in this chapter. Potential errors in the use of fluorescent probes to determine binding parameters have been discussed by Zierler (1977). Fortunately, sufficient quantities of cholinergic receptor are available to make such studies possible. This is, in general, not the case for receptors for polypeptide hormones. For example, the studies with fluorescent cholinergic probes employed receptor concentrations of about 2.5×10^{-7} M; it can be calculated that an equivalent experiment for the study of insulin receptors in fat cells would require the impossible concentration of about 10^{10} cells ml^{-1}. From the known fluorescence emission of a particular ligand derivative and from an estimate of the number of receptors present in a given soluble receptor preparation, it should be possible to estimate the feasibility of such studies.

(i) *Differential adsorption of ligand*

The differential precipitation or adsorption (e.g. by ion exchange or by lectin–agarose immobilization) of the hormone receptor complex was discussed above. Alternatively, the differential rapid removal of the free ligand from solution provides another approach for the study of soluble receptors. For example, charcoal or talc adsorption has been useful for radioimmunoassay procedures to separate free from antibody-bound ligand; a similar approach has yielded good results for the assay of estrogen-binding macromolecules in tissue extracts (Korenman, 1970, 1975). Affinity chromatographic techniques provide another means of selectively removing free ligand, for instance, with ligand-specific antibody attached to an inert support. It should be pointed out, however, that the use of antisera against cobra neurotoxin to precipitate free toxin, but not the toxin–receptor complex, was limited to the assay of relatively crude receptor solutions; the purified receptor–toxin complex was coprecipitated with free toxin (Klett *et al.*, 1973, 1974).

It is evident that several approaches are possible for the study of solubilized receptors. While each receptor may present its idiosyncratic difficulties for the assay of binding, one or other approach should provide a means of assay. Irrespective of the method finally chosen, the same criteria for evaluating the binding of ligands to membrane-associated receptors should be used to assess studies with soluble preparations.

1.3.3 Other methods for receptor detection

In addition to the methods for measuring ligand binding discussed above, a number of physical methods (e.g. use of intrinsic membrane fluorescence or incorporation into membranes of either fluorescent or spin-label probes) have been used to detect ligand—receptor interactions. For instance, Sonenberg and co-workers have observed marked changes in membrane fluorescence in the presence of growth hormone (Sonenberg, 1969, 1971; Rubin *et al.*, 1973a,b; Aizono *et al.*, 1974; Postel-Vinay *et al.*, 1974a,b). Electron spin resonance measurements have been of use in studying prostaglandin E and E_2 interaction with erythrocytes (Kurtz *et al.*, 1974) and in evaluating the binding of spin-labeled acetylcholine derivatives (Brisson *et al.*, 1975). Despite the success of those studies, these methods have not been widely applied in the study of receptor—ligand interactions. As illustrated in Chapter 8, ultrastructural methods have also yielded considerable information about membrane-localized ligand binding sites. Although it is not always a simple task to prepare a suitable probe for such studies (e.g. ferritin- or lactoperoxidase-conjugated ligands), the ultrastructural information gained by such studies (e.g. localization of cell type or cellular polarity of binding) adds substantially to measurements of the overall binding of ligand by a preparation of interest. Electron microscopic autoradiography can serve not only to localize ligand binding within the plane of the membrane (Jarett and Smith, 1974; Gorden *et al.*, 1978) but to quantitate the amount of ligand bound per cell. For instance, the estimates *in vivo* of insulin binding to hepatocytes by quantitative electron microscopic methods (Bergeron *et al.*, 1977) are in quite close agreement with measurements of hepatocyte insulin binding by other methods. In general, in ultrastructural studies, compared with other ligand binding methods, it is more difficult to establish the ligand specificity and other criteria that are required of a 'true' ligand—receptor interaction.

1.4 USE OF MEMBRANE RECEPTORS FOR LIGAND DETECTION AND QUANTITATION

In ligand binding assays, information is obtained not only about the receptor but also about the ligand bound. Because of the high affinity, specificity and ready availability of membrane receptors, such preparations provide a suitable alternative to antibodies for ligand assays (so-called 'radioreceptor' assays). Because in theory, radioreceptor assays measure only biologically active constituents, the information provided is complementary to immunoassay data. In practice, however, radio-receptor assays have proved of only limited usefulness compared with the more routinely used immunoassay. Nonetheless, because of the importance of a radio-receptor assay in certain settings (e.g. for analysis of gonadotrophic (Catt *et al.*, 1976) or lactogenic hormones (Shiu and Friesen, 1976), a number of factors that relate specifically to 'radioreceptor' assays will be explored in the following sections.

As for any other protein binding assay (for a survey and introduction to this field, see Buttner *et al.*, 1976; Thorell and Larsen, 1978) the following items require evaluation:

(1) preparation of ligand binder (receptor);

(2) preparation of radiolabeled ligand tracer and standard solutions;

(3) optimization of conditions for ligand binding and for separation of free from bound ligand; and

(4) standardization of the assay for routine use.

In radioreceptor assays, the concerns regarding the preparation of labeled tracer and the separation of free from bound ligand are as discussed in previous sections. Membrane preparations suitable for radioreceptor assays can usually be obtained in quantity from a recognized target tissue for the ligand of interest. For instance, rabbit breast tissue may serve as a source of membranes for the assay of prolactin (Shiu and Friesen, 1976); testis membranes can be used for the assay of gonadotropins (Catt *et al.*, 1976) and adrenocortical membranes may prove of use for ACTH assays (Finn and Hofmann, 1976). Since membranes from placenta (usually human) possess receptors for a large number of hormones (for example, insulin, epidermal growth factor—urogastrone, catecholamines), a convenient method for preparation of membranes from human placenta is described.

1.4.1 Preparation of membranes from placenta for radioreceptor assay

Membranes from human placenta are obtained after homogenization by differential centrifugation according to a modification of a method originally used for rat liver membranes (Cuatrecasas, 1972c). Fresh frozen human placenta is thawed overnight at 4°C in 0.25 M sucrose buffered with 25 mM Tris—HCl, pH 7.4, dissected free of chorion, amnion, large vessels and umbilical cord, and cut into small pieces. Washed tissue (approximately 300 g) is resuspended in 10 volumes (w/v) of the sucrose—Tris solution and homogenized at 0°C with a rotary knife homogenizer (Brinkmann Polytron, 90 s, setting 7). The homogenate is filtered through two layers of cheesecloth and centrifuged at 600 g for 10 min at 4°C; the supernatant is then centrifuged at 10 000 g for 30 min. These procedures can be performed in large plastic bottles. The supernatant and the light pink fluffy top portion of the pellet are decanted, made 0.1 M in NaCl and 0.2 mM in $MgSO_4$ and then centrifuged at 49 000 g for 40 min at 4°C. The pellet is resuspended in 50 mM Tris—HCl pH 7.4 and centrifuged again at 48 000 g for 30 min at 4°C. The resulting pellet is resuspended by gentle homogenization in 20 ml of the Tris buffer to yield a suspension of about 2—10 mg ml^{-1} protein. The suspension can be stored in aliquots at −20°C for periods of up to at least one year and thawed as required for use [usually about 50 μg protein per assay tube (Nexø *et al.*, 1977)]. This preparation possesses large numbers of specific high affinity binding sites for insulin and epidermal growth factor—urogastrone and is particularly useful for monitoring the quality of preparations of iodinated derivatives of these two polypeptides.

1.4.2 Special features of radioreceptor assays

In contrast to radioimmunoassays, factors such as pH, ionic strength, divalent cation concentration, buffer composition and temperature are of critical importance in radioreceptor assays and must be evaluated for each ligand studied. In addition to the ability of receptors to distinguish between bioactive and inactive ligands, receptors, unlike antibodies, are able to detect equiactive ligands, such as mouse and human epidermal growth factor—urogastrone, that have quite different amino acid sequences, so as not to be immunologically cross-reactive. This 'degeneracy' of radioreceptor assays can be an advantage, in that the same assay can be used to measure analogous (but not chemically identical) ligands from several species.

The receptor affinity is an important factor in radioreceptor assays, since the sensitivity of the assay relates directly to the dissociation constant (K_d) of the receptor. Reliable sensitive measurements can only be done if the K_d is equal to or lower than the concentration of the substance to be measured. For instance, the radioreceptor assay for EGF—URO (Nexø *et al.*, 1977) uses membranes with a K_d of about 10^{-9} M; this assay is ideal for measuring EGF—URO in urine samples which possess concentrations of about 10^{-8} M. Suitable dilutions of urine can be prepared to avoid complications in the assay due to solutes (salt, proteolytic enzymes) other than the ligand of interest. In contrast, the use of this assay to measure serum EGF—URO concentrations (less than 10^{-9} M) is limited by the sensitivity of the assay. In practice, the somewhat low receptor affinity, compared with the high ligand affinity of antibody is a limiting factor in the routine use of radioreceptor assay.

1.4.3 Avoiding artifacts in radioreceptor assays

In using radioreceptor assays for the measurement of ligands in biological fluids, it is important to determine that the reduced binding of radioactive ligand caused by the test sample is really due to the presence of active ligand in the unknown and is not caused by other interfering processes. Firstly, it is important to demonstrate that binding competition by the unknown is truly competitive with the radiolabeled ligand. In using dilutions of the unknown, it should be possible to demonstrate binding competition curves that are parallel with standard curves generated with pure, unlabeled ligand. Data analysis [e.g. according to Scatchard (1949)] should also indicate competitive binding kinetics. In addition, it should be possible to demonstrate reversible binding inhibition by the unknown. This is particularly important when unknown samples are thought to contain proteolytic enzymes which might destroy the receptor so as to yield artifactually-reduced ligand binding and falsely elevated estimates of ligand concentrations. Receptor destruction can be readily assessed for receptor preparations in which the bound ligand can be

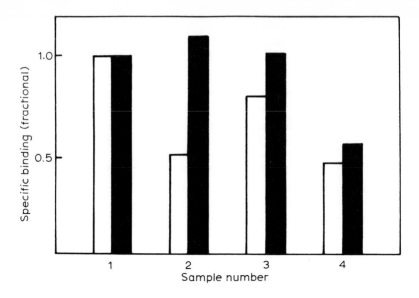

Fig. 1.5 A test for receptor damage in radioreceptor assays. The fractional
specific binding of radiolabeled mouse EGF—URO (ordinate) is expressed
relative to the binding observed in the absence of competing agent. Open
bars: fractional binding in the presence of test samples. (1) Buffer alone,
(2) buffer containing 8 pM mouse EGF—URO, (3) human urine sample A
and (4) human urine sample B. Closed bars: Membranes incubated with the
above samples were regenerated by the acid wash, as described in the text,
and were again tested for fractional binding in the presence of buffer alone.
Full binding activity is regained for samples No. 2 and 3; sample 4, however,
causes a reduction in the regenerated binding activity, indicating the presence
of a receptor-destroying substance in urine sample B.

efficiently dissociated without receptor destruction. A test for receptor destruction
is illustrated in Fig. 1.5. The test makes use of the observation that the placenta
receptor for EGF—URO can be regenerated after ligand binding by incubation in
50 mM sodium acetate buffer pH 4.0. Thus, both pure EGF—URO and an apparently
equivalent amount of unknown can be equilibrated with membranes which are then
harvested in a filter manifold on EGWP Millipore filters (0.2 μm pore size) as
described above. With the filters still in place, harvested membranes are regenerated
by washing with three 3 ml aliquots (two aliquots filtered under vacuum; the third
filtered by gravity over about 10 min) of 50 mM acetate buffer, pH 4.0 followed
by a wash (10 ml) with the binding assay buffer at pH 7.4. Radiolabeled EGF—URO
(in 0.3 ml) is then equilibrated on top of the filters (30 min at 24°C) and the
membranes are finally washed with ice-cold buffer under reduced pressure as for a
routine binding assay. The binding of radioligand to untreated membranes and to

membranes incubated with test samples can be compared before (open bars, Fig. 1.5) and after (closed bars, Fig. 1.5) the regeneration procedure. In control experiments, virtually all of the receptor binding capacity of the membranes can be regenerated by the acid wash; receptor destruction is indicated by a reduction in the binding of radioligand by the immobilized regenerated membranes as illustrated by sample No. 4, Fig. 1.5.

Another artifact in membrane radioreceptor assays may be caused by the presence in unknown samples of soluble substances that can interact with the radiolabeled probe. Should this be the case, the routine binding assay in which membranes, radiolabeled ligand and unknown are incubated together, would yield spuriously low values of radioligand binding because the tracer would be removed with the soluble binding substance during the membrane washing procedure. The ligand affinity of the competing soluble substance need not be high, provided it is present in high concentrations in the test solution, relative to the receptor concentration. The occurrence of this possible complication can be assessed by sequential incubation of the membrane receptor first with the unknown solution (or standard concentrations of unlabeled ligands) and then with the radiolabeled ligand. The initial incubation should be done with sufficient amounts of both unknown and unlabeled ligand to achieve substantial 'saturation' of the receptor. The membranes can then be harvested on millipore filters in a filter manifold, rinsed rapidly with ice-cold buffer and equilibrated with radiolabeled ligand (15 min at 24°C) while still in the manifold. Using this method, binding competition by unlabeled EGF–URO can be readily detected using placenta membranes. This approach was of particular value in a study of bovine milk samples which were negative in a radioimmunoassay but which appeared to contain an EGF–URO-like substance according to the routine radio-receptor assay. However, the milk samples yield negative results by the above sequential procedure. We were thus led to the conclusion that bovine milk, which does not exhibit EGF–URO receptor destroying activity by the regeneration test described above, probably contains a soluble substance that can compete for the receptor binding of EGF–URO. The sequential incubation method would also reveal the inhibitory effect of other solutes in the unknown solution (e.g. calcium chelating substances) that might inhibit the binding of radioligand but would not irreversibly destroy the receptor.

1.4.4 Role for radioreceptor assays

In general, radioreceptor assays do not provide either the sensitivity or the precision of the radioimmunoassay. Nonetheless, receptor preparations are more readily obtained than are antibodies and in contrast with the immunoassay, the radio-receptor assay measures only the biologically active ligand. Thus, radioreceptor assays can be conveniently developed for ligands lacking readily available antibody (e.g. human EGF–URO) and can yield information that is complementary to immunoassay data. A particularly useful property of the radioreceptor assay relates to the detection of biologically active ligands that may differ structurally from the radiolabeled ligand probe. A striking use of this property can be seen in the detection and isolation of endogenous brain 'opiate'-like peptides using a brain membrane receptor preparation and radiolabeled opiates (Simantov *et al.*, 1977). Further, a radioreceptor assay has proved enormously helpful in the study of human chorionic gonadotrophin (Saxena *et al.*, 1974), of lactogenic hormones (Shiu and Friesen, 1976), and in the isolation from RNA tumor virus-transformed cells of a factor that interacts with the receptor for EGF—URO (DeLarco and Todaro, 1978). Thus, although radioreceptor assays may not prove of widespread clinical usefulness, they provide an invaluable research tool. Moreover, although the assays *per se* do not actually lead to further information about the receptor, their use may lead to the discovery of endogenous factors that may interact with the receptor in a manner different from that of the original radiolabeled ligand probe used.

1.5 CONCLUSION

From the above discussion, it is evident that a variety of methods have been developed to measure ligand binding interactions for both particulate and soluble receptors. On the one hand, one's focus may be on the receptor characteristics, with the goal of receptor purification in mind. Alternatively, the binding assays may be used for quantitative estimates of the ligand bound and for the purification of endogenous ligands that may be structurally dissimilar from the radioligand probe initially studied. The assays to date have been a cornerstone in the study of hormone receptors and in the future ought to yield much interesting information about other analogous binding proteins.

REFERENCES

Aizono, Y., Roberts, J.E., Sonenberg, M. and Swislocki, N.I. (1974), *Arch. Biochem. Biophys.,* **163**, 634.

Atlas, D., Steer, M.L. and Levitzki, A. (1974), *Proc. natn. Acad. Sci. U.S.A.,* **71**, 4246.

Bergeron, J.J.M., Levine, G., Sikstrom, R., O'Shaughnessy, D., Kopriwa, B., Nadler, N.J. and Posner, B.I. (1977), *Proc. natn. Acad. Sci. U.S.A.,* **74**, 5051.

Brisson, A.D., Scandella, C.J., Bienvenue, A., Devaux, P.F., Cohen, J.B. and Changeux, J-P. (1975), *Proc. natn. Acad. Sci., U.S.A.,* **72**, 1087.

Buttner, J., Borth, R., Boutwell, J.H., Broughton, P.M.G. and Bowyer, R.C. (1976), *Clin. Chem. clin. Biochem.,* **14**, 265.

Catt, K.J., Ketelslegers, J.M. and Dufau, M.L. (1976), In: *Methods in Molec. Biology,* Volume 9, Part 1 (Blecher, M., ed.), p. 175, Marcel Dekker, New York.

Chang, K-J, and Cuatrecasas, P. (1974), *J. biol. Chem.,* **249**, 3170.

Cheng, Y-C. and Prusoff, W.H. (1973), *Biochem. Pharmacol.*, **22**, 3099.
Chesebro, B. and Svehag, S.E. (1968), *Clin. Chim. Acta*, **20**, 527.
Clark, A.J. (1926a), *J. Physiol.*, (London) 61, 530.
Clark, A.J. (1926b), *J. Physiol.*, (London) 61, 547.
Clark, A.J. (1933), *Mode of Action of Drugs on Cells*, Arnold, London.
Cohen, J.B. and Changeux, J.-P. (1973), *Biochemistry*, **12**, 4855.
Cuatrecasas, P. (1972a), *J. biol. Chem.*, **247**, 1980.
Cuatrecasas, P. (1972b), *Proc. natn. Acad. Sci.*, *U.S.A.*, **69**, 318.
Cuatrecasas, P. (1972c), *Proc. natn. Acad. Sci.*, *U.S.A.*, **69**, 1277.
Cuatrecasas, P. and Hollenberg, M.D. (1975), *Biochem. biophys. Res. Commun.*, **62**, 31.
Cuatrecasas, P. and Hollenberg, M.D. (1976), *Adv. prot. Chem.*, **30**, 251.
Cuatrecasas, P., Fuchs, S. and Anfinsen, C.B. (1967), *J. biol. Chem.*, **242**, 3063.
Cuatrecasas, P., Hollenberg, M.D., Chang, K.J. and Bennett, V. (1975), *Recent Progr. Horm. Res.*, **31**, 37.
DeLarco, J.E. and Todaro, G.J. (1978), *Proc. natn. Acad. Sci.*, *U.S.A.*, **75**, 4001.
DeRobertis, E. (1971) *Science*, **171**, 963.
Desbuquois, B. and Aurbach, G.D. (1971), *J. clin. Endocrinol. Metab.*, **33**, 732.
Eldefrawi, M.E. and Eldefrawi, A.T. (1973), *Archs. Biochem. Biophys.*, **159**, 362.
Finn, F.M. and Hofmann, K. (1976), In: *Methods in Molecular Biology*, Volume 9, Part 1, (Blecher, M., ed.), p. 37, Marcel Dekker, New York.
Franklin, G.I. and Potter, L.T. (1972), *FEBS Letters*, **28**, 101.
Giorgio, N.A., Johnson, C.B. and Blecher, M. (1974), *J. biol. Chem.*, **249**, 428.
Gliemann, J., Osterlind, K., Vinten, J. and Gammeltoft, S. (1972), *Biochem. biophys. Acta*, **286**, 1.
Gorden, P., Carpentier, J.-L., Freychet, P., LeCam, A. and Orci, L. (1978), *Science*, **200**, 782.
Haga, T., Haga, K. and Gilman, A. (1977), *J. biol. Chem.*, **252**, 5776.
Hebert, T.T. (1963), *Phytopathology*, **53**, 362.
Henderson, R. and Wang, J.H. (1972), *Biochemistry*, **11**, 4565.
Hock, R.A., Nexϕ, E. and Hollenberg, M.D. (1979a), *Nature*, **277**, 403.
Hock, R.A., Nexϕ, E. and Hollenberg, M.D. (1979b), *J. biol. Chem.*, submitted.
Hollenberg, M.D. and Cuatrecasas, P. (1979), In: *The Receptors, a Comprehensive Treatise*, Vol. 1, (O'Brien, R.D., ed.), p. 193, Plenum, New York.
Hummel, J.P. and Dreyer, W.J. (1962), *Biochem. Biophys. Acta*, **63**, 530.
Jacobs, S., Chang, K. and Cuatrecasas, P. (1975), *Biochem. biophys. Res. Commun.*, **66**, 687.
Jarett, L. and Smith, R.M. (1974), *J. biol. Chem.*, **249**, 7024.
Juckes, I.R.M. (1972), *Biochem. Biophys. Acta*, **229**, 535.
Klett, R., Fulpius, B.W., Cooper, D., Smith, M., Reich, E. and Possani, L.D. (1973), *J. biol. Chem.*, **248**, 6841.
Klett, R., Fulpius, B.W., Cooper, D. and Reich, E. (1974), In: *Synaptic Transmission and Neuronal Interaction*, (Bennett, M.V.L., ed.), p. 179, Raven, New York.
Korenman, S.G. (1970), *Endocrinology*, **87**, 1119.
Korenman, S.G. (1975), In: *Methods in Enzymology*, Volume 36, Part A, (O'Malley, B.W. and Hardman, J.G., eds.), p. 49, Academic Press, New York.

Krupp, M.N. and Livingston, J.N. (1978), *Proc. natn. Acad. Sci., U.S.A.*, **75**, 2593.

Krupp, M.N. and Livingston, J.N. (1979), *Nature*, **278**, 61.

Kurz, P.G., Ramwell, P.W. and McConnell, H.M. (1974), *Biochem. biophys. Res. Commun.*, **56**, 478.

Kushner, L.M. and Hubbard, W.D. (1954), *J. Phys. Chem.*, **58**, 1163.

Levitzki, A., Atlas, D. and Steer, M.L. (1974), *Proc. natn. Acad. Sci., U.S.A.*, **71**, 2773.

Levitzki, A., Sevilla, N., Atlas, D. and Steer, M.L. (1975), *J. molec. Biol.*, **97**, 35.

Livingston, J.N., Cuatrecasas, P. and Lockwood, D.H. (1974), *J. Lipid Res.*, **15**, 26.

March, S.G., Parikh, I. and Cuatrecasas, P. (1974), *Analyt. Biochem.*, **60**, 149.

Martinez-Carrion, M. and Raftery, M.A. (1973), *Biochem. biophys. Res. Commun.*, **55**, 1156.

Mayer, R. and Himmel, C. (1972), *Biochemistry*, **11**, 2082.

Meunier, J.C., Olsen, R.W. and Changeux, J.P. (1972a), *FEBS Letters*, **64**, 63.

Meunier, J.C., Olsen, R.W., Menez, A., Fromageot, P., Boquet, P. and Changeux, J.-P. (1972b), *Biochemistry*, **11**, 1200.

Moody, T., Schmidt, J. and Raftery, M.A. (1973), *Biochem. biophys. Res. Commun.*, **53**, 761.

Mooser, G., Schulman, H. and Sigman, D. (1972), *Biochemistry*, **11**, 1595.

Nexø, E., Hock, R.A. and Hollenberg, M.D. (1979) *J. biol. Chem.*, **254**, 8740.

Nexø, E. and Hollenberg, M.D. (1980), *Biochem. biophys. Acta*, **628**, 190.

Nexø, E., Nelson, J., Lamberg, S.I. and Hollenberg, M.D. (1977), *Clin. Res.*, **25**, 657A.

Polson, A. and Deeks, D. (1963), *J. Hyg.*, **61**, 149.

Polson, A., Potgieter, G.M., Largier, J.F., Mears, G.E.F. and Joubert, F.J. (1964), *Biochim. Biophys. Acta*, **82**, 463.

Postel-Vinay, M.C., Sonenberg, M. and Swislocki, N.I. (1974a), *Biochim. Biophys. Acta*, **332**, 156.

Postel-Vinay, M.C., Swislocki, N.I. and Sonenberg, M. (1974b), *Endocrinology*, **95**, 1554.

Radda, G.K. and Van der Kooi, J. (1972), *Biochim. Biophys. Acta*, **265**, 509.

Rubin, M.S., Swislocki, N.I. and Sonenberg, M. (1973a), *Archs. Biochem. Biophys.*, **157**, 243.

Rubin, M.S., Swislocki, N.I. and Sonenberg, M. (1973b), *Archs. Biochem. Biophys.*, **157**, 252.

Sahyoun, N., Hollenberg, M.D., Bennett, V. and Cuatrecasas, P. (1977), *Proc. natn. Acad. Sci., U.S.A.*, **74**, 2860.

Saxena, B.B., Hasan, S.H., Haour, F. and Schmidt-Gallwitzer, M. (1974), *Science*, **184**, 793.

Scatchard, G. (1949), *Ann. N.Y. Acad. Sci.*, **51**, 660.

Schmidt, J. and Raftery, M.A. (1973a), *Biochemistry*, **12**, 852.

Schmidt, J. and Raftery, M.A. (1973b), *Anal. Biochem.*, **52**, 349.

Seligman, P.A. and Allen, R.H. (1978), *J. biol. Chem.*, **253**, 1766.

Shiu, R.P.C. and Friesen, H.G. (1976), In: *Methods in Molecular Biology*, Volume 9, Part 2, p. 565, Marcel Dekker, New York.

Sica, V., Nola, E., Parikh, I., Puca, G.A. and Cuatrecasas, P. (1973a), *Nature New Biol.*, **244**, 36.

Sica, V., Parikh, I., Nola, E., Puca, G.A. and Cuatrecasas, P. (1973b), *J. biol. Chem.*, **248**, 6543.

Silhavy, T.J., Szmeleman, S., Boos, W. and Schwartz, M. (1975), *Proc. natn. Acad. Sci., U.S.A.,* **72**, 2120.

Simantov, R., Childers, S.R. and Snyder, S.H. (1977), *Brain Res.,* **135**, 358.

Snyder, S.H., Pasternak, G.W. and Pert, C.B. (1975), In: *Handbook of Psychopharmacology,* Volume 5, p. 329, (Iversen, L.L., Iversen, S.D. and Snyder, S.H., eds.), Plenum, New York.

Sonenberg, M. (1969), *Biochem. biophys. Res. Commun.,* **36**, 450.

Sonenberg, M. (1971), *Proc. natn. Acad. Sci., U.S.A.,* **68**, 1051.

Straus, O.H. and Goldstein, A.J. (1943), *J. gen. Physiol.,* **26**, 559.

Thorell, J.I. and Larson, S.K. (1978), *Radioimmunoassay and Related Techniques.* The C.V. Mosby Company, St. Louis.

Waggoner, A. and Stryer, L. (1970), *Proc. natn. Acad. Sci., U.S.A.,* **67**, 579.

Weber, G., Borris, D., DeRoberts, E., Barrantes, F., LaTorre, J. and DeCarlin, M. (1971), *Mol. Pharmacol.,* **7**, 530.

Zierler, K. (1977), *Biophys. Struct. Mechanism,* **3**, 275.

2 Solubilization and Characterization of Membrane Proteins

JACQUELINE A. REYNOLDS

Acknowledgements
The author is particularly grateful to graduate students enrolled in a course in Biochemistry of Membranes for helping to clarify many of the ideas expressed here and to Dr Charles Tanford for many helpful discussions. In addition, thanks are due to both Balliol College, Oxford, for providing facilities during a Sabbatical Leave as a J.S. Guggenheim Fellow and to the Department of Chemistry, University of Washington, where most of this manuscript was actually written. Appreciation is particularly extended to the John Simon Guggenheim Foundation for financial support during the time this work was in progress.

Membrane Receptors: Methods for Purification and Characterization
(*Receptors and Recognition,* Series B, Volume 11)
Edited by S. Jacobs and P. Cuatrecasas
Published in 1981 by Chapman and Hall, 11 New Fetter Lane, London EC4P 4EE
© Chapman and Hall

2.1 INTRODUCTION

A major advance in the study of membrane-bound proteins was made by Henderson and Unwin when they published a 7 Å resolution map of the purple membrane from *Halobacterium halobium*. The unique feature of this membrane is that its major protein, bacteriorhodopsin, forms a two-dimensional crystal within the plane of the bilayer thus making it possible to apply electron microscopy and diffraction techniques to this system. Most membranes do not form two-dimensional crystalline arrays, however, and other methods must be invoked for studying the functioning proteins associated with lipid bilayers.

In the past few years significant progress has been made in studying membrane-bound proteins at the molecular level, primarily through the use of detergents as ligands to stabilize the native conformation in a solubilized state (Helenius and Simons, 1975; Tanford and Reynolds, 1976). It is not my intention to provide a detailed review of the literature in this field but rather to select examples of solubilization, purification and characterization of membrane proteins. These examples will hopefully demonstrate what I believe to be a rational approach to the problem in general and will emphasize experimental procedures designed to provide maximal information.

Unfortunately, we are not yet in a position to generalize widely as to which particular type of detergent is most suitable for a specific application. The vast amount of literature relating to this problem cannot be dealt with in entirety within these spatial confines but I will attempt to summarize fundamental principles and the few systematic studies available. This chapter, then, is essentially a report on the 'state of the art'.

2.2 PROTEINS IN MEMBRANE-BOUND FORM

It has become common usage to divide membrane proteins into two operational classifications. One type of membrane protein (extrinsic) is defined as that which has aqueous solubility in the absence of bound amphiphile, but is found *in vivo* attached to the surface of a biological membrane through specific interactions with lipids or other membrane proteins (Green, 1972; Singer and Nicolson, 1972). The second type of membrane protein is in intimate contact with the hydrocarbon portion of the lipid bilayer and cannot be solubilized into an aqueous medium without the use of substitute amphiphilic ligands such as detergents. This latter classification of proteins is often referred to as intrinsic or integral and it is this group with which this chapter is concerned.

Intrinsic membrane proteins are associated *in vivo* with both a hydrophobic environment and an aqueous solution, that is, they consist of domains of differing solvent affinities. The structure and function of these proteins are dependent upon interactions of each of these domains with other chemical components, some of which are in the interior of the lipid bilayer and others which are at the surface or in the aqueous medium. Thus, these intrinsic membrane proteins live in an asymmetric world. Not only are lipids arranged asymmetrically in the membrane (Bretscher, 1973), but the aqueous environment differs on the internal and external faces. An electrical potential exists across most membranes thus subjecting the proteins to a field gradient.

In Section 2.3 the problem of duplicating the membrane environment by the use of detergents is discussed in some detail. However, it is obvious that the asymmetry of an intact membrane cannot be duplicated in a one-compartment system consisting of soluble detergent—protein complexes. Thus, it would not be surprising to find some functional differences between these two systems. If meaningful information is to be obtained from studies of solubilized forms of intrinsic membrane proteins, we need to know exactly what those differences are and attempt to relate them to the environmental conditions extant in each system. Complete functional characterization of the protein in the intact membrane is then the first step in a study of intrinsic membrane proteins and is used as the reference point for comparison with a solubilized form of the same protein. A particularly lucid discussion of essential functional characterization of hormone receptors prior to solubilization and purification is presented by Cuatrecasas and Hollenberg (1976). These authors emphasize especially the need to study the membrane-bound form of proteins in the intact cell and not only in partially purified membrane fractions.

2.3 PROPERTIES OF DETERGENTS

The recent literature contains detailed discussions of the theory of micelle formation (Tanford, 1973, 1974) and descriptions of detergent properties (Helenius and Simons, 1975; Steele *et al.*, 1978; Helenius *et al.,* 1979). In this section the aspects of detergent behavior in aqueous solutions that are of particular importance in their use as membrane protein solubilizing agents and as ligands for membrane proteins are summarized.

Detergents are amphiphilic molecules containing spatially separated hydrophobic and hydrophilic regions. In aqueous solution these compounds form monolayers at interfaces and self-associated structures in the bulk solution called micelles. The principal driving force for micelle and monolayer formation is the hydrophobic free energy which arises from the unfavorable effect of the hydrophobic regions of amphiphiles on water structure. The concentration at which micelle formation occurs is called the critical micelle concentration (CMC), and the thermodynamic equilibrium involved can be satisfactorily described by the following equation.

$$mL \rightleftharpoons L_m ; K = \frac{[L_m]}{[L]^m} \qquad (2.1)$$

It is apparent from inspection of equation 2.1 that if m is large, $[L]$, the concentration of monomer in solution, will increase very little above the CMC, i.e. when $[L_m]$ becomes a significant fraction of the total concentration.

The average aggregation number, $\bar{m}$, and the CMC of a detergent containing an ionic head group are functions of the ionic strength of the solution, the former increasing and the latter decreasing as the ionic strength is raised. Detergents with neutral polar head groups are relatively insensitive to ionic strength, and thus the size of the micelle and the CMC cannot be manipulated by altering the solution environment.

Commercial detergents of a wide variety are available, and the most common hydrophobic and hydrophilic groups which they contain are shown in Table 2.1. The reader is referred to more comprehensive listings which include trade names (Rosen and Goldsmith, 1972; Steele *et al.*, 1978; Helenius *et al.*, 1979).

Micelles formed from amphiphiles such as those shown in Table 2.1 usually form small oblate ellipsoids (Tanford, 1973). Fig. 2.1 is a schematic diagram showing two different types of micelles together with a typical phospholipid bilayer. The hydrophobic regions of all three amphiphiles have been removed from contact with water through the process of self-association and the dimensions of the hydrocarbon interior are determined by the length of the hydrophobic moiety. It is obvious that one can mimic the hydrophobic dimensions of a phospholipid bilayer by choosing a detergent containing a normal alkyl or alkenyl chain of appropriate length corresponding to that of the fatty acyl chains in a naturally occurring lipid.

The hydrophilic head groups of detergents are hydrated and vary considerably in length of extension from the hydrophobic-water interface. The polyoxyethylenes in particular are quite bulky and one can envision the possibility of steric problems in their interaction with some membrane proteins.

A totally different type of amphiphile from those shown in Table 2.1 and Fig. 2.1 is found in the bile salts. These compounds contain a fused ring system with hydrophobic and hydrophilic *faces*. They aggregate in small clusters and also undergo secondary aggregation (Fig. 2.2). Their CMCs are highly pH dependent as are their aggregation numbers (Small, 1971). This latter property makes them somewhat undesirable for applications discussed in this chapter since at neutral pH these compounds are not totally ionized and tend to form very large, ill-behaved micelles. In addition, one is uncomfortable about substituting these rigid hydrophobic moieties for the normally fluid hydrocarbon chains found on phospholipids.

All amphiphilic compounds, including lipids, establish an equilibrium between monomer and micelle in aqueous solution, and it is important for the discussion in subsequent sections to keep in mind that this is a reversible thermodynamic process. The kinetics of the association–dissociation process are also involved in biological applications. The association constant, K in equation (2.1) is the quotient of two rate constants, k_1 (forward) and k_2 (reverse). As the CMC decreases

Table 2.1 Hydrophobic and hydrophilic groups most frequently found in commercial detergents

Hydrophobic tails	Hydrophilic heads

1. $CH_3 (CH_2)_n -$

2. $CH_3 (CH) = (CH_2)_n -$

3.
$$CH_3 - C - CH_2 - C - \langle\rangle -$$
with CH_2 groups branching

4.
$$-O-\overset{\overset{O}{\|}}{\underset{\underset{O}{}}{S}}-O^-$$

5.
$$-\overset{\overset{CH_3}{|}}{\underset{\underset{CH_3}{|}}{N_+}}-CH_3$$

6.
$$-\overset{\overset{CH_3}{|}}{\underset{\underset{CH_3}{|}}{N_+}}-CH_2-CH_2-\overset{\overset{O}{\|}}{C}-O^-$$

7. $-O-(CH_2-CH_2-O)_x H$

Triton Series – 3, 7
Lubrol and Brij Series – 1, 2, 7
Alkyl trimethylammonium – 1, 5

Betaines – 1, 6
Alkyl sulfates – 1, 4

Tween Series – 1, 2, 8
Lysolipids – 1, 9

(continued on the next page)

Table 2.1 Hydrophobic and hydrophilic groups most frequently found in commercial detergents (*continued*)

Hydrophilic heads

8.

$$HO-(CH_2-CH_2-O)_w-HC \underset{\displaystyle \overset{\displaystyle H_2C}{\big|}}{\overset{\displaystyle O}{\diagdown}} \begin{array}{l} CH-CH-CH_2-(O-CH_2-CH_2)_x-O-C- \\ | \\ CH(O-CH_2-CH_2)_y-OH \\ | \\ (O-CH_2-CH_2)_z-OH \end{array} \overset{\displaystyle O}{\overset{\displaystyle \|}{}}$$

9. $$\begin{array}{l} -O-CH_2 \\ | \\ HO-CH \quad \overset{O^-}{\big|} \\ | \quad \quad \| \\ CH_2-O-P-O-R \\ | \\ O \end{array}$$

Triton Series – 3, 7   Betaines – 1, 6   Tween Series – 1, 2, 8
Lubrol and Brij Series – 1, 2, 7   Alkyl sulfates – 1, 4   Lysolipids – 1,9
Alkyl trimethylammonium – 1, 5

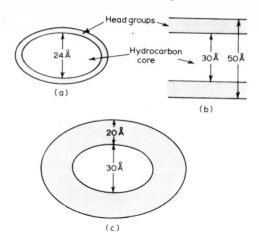

Fig. 2.1 Approximate dimensions of disk-shaped micelles in comparison with a planar bilayer. (a) SDS micelle at ionic strength 0.1. (b) Bilayer formed by egg yolk phosphatidyl choline. (c) Lubrol WX micelle. The thickness of the hydrophobic core is determined solely by the alkyl chain length. The shaded areas reflect the distances to which the head groups extend from the core surface. The head groups themselves do not fill all the space within these regions, space between head groups being occupied by solvent. (Reproduced with permission of C. Tanford, J.A. Reynolds and Elsevier Scientific Publishing Company, Amsterdam.)

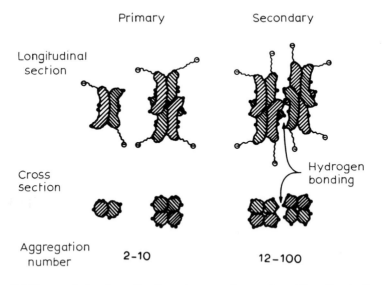

Fig. 2.2 Proposed structures for the primary and secondary bile salt micelles. (Reproduced with permission of A. Helenius, K. Simons, and Elsevier Scientific Publishing Company, Amsterdam.)

(increasing K), the reverse rate constant becomes smaller resulting in slower rates of dissociation of monomer from micelle. Experimental implications of this will be discussed further in Section 2.3.1.

Some non-ioniç detergents such as the polyoxyethylenes undergo a secondary aggregation process (Helenius and Simons, 1975; Tanford and Reynolds, 1976) which is dependent upon the relative lengths of the hydrophobic and hydrophilic regions. This can be avoided by increasing the length of the polyoxyethylene chain relative to the alkyl chain. For example, $C_{12}E_6$ forms large aggregates at room temperature, but $C_{12}E_8$ maintains a small micelle size at high concentrations and at the same temperature. For the purposes of physical characterization of protein–detergent complexes secondary aggregation phenomena must be avoided.

2.3.1 Solubilization and purification in detergents

(a) *Removal of extrinsic membrane proteins*
All membranes contain extrinsic proteins which are associated with the surface of the bilayer. The most well-known examples are the cytoskeletal elements such as actin, tubulin, etc. (Korn, 1978). These proteins can usually be solubilized without seriously perturbing the lipid bilayer by varying the ionic strength or by adding appropriate chelating agents. Spectrin, for example, is a filamentous protein located on the cytoplasmic surface of the human erythrocyte membrane. It is released from association with the bilayer by reducing the concentration of inorganic cations in the solution (Marchesi and Palade, 1967; Reynolds and Trayer, 1971). The resultant membrane is fragmented but still morphologically intact as a bilayer.

Another example of removal of extrinsic proteins is shown in Fig. 2.3a and b which are sodium dodecyl sulfate polyacrylamide gels of garfish olfactory axolemma prepared in 0.1 M $NaHCO_3$, pH 8.0, and subsequently washed with EDTA (Nielsen, 1977). One intrinsic membrane protein which is present in this system is a Na^+, K^+-ATPase which is retained in the membrane after the EDTA wash. The arrow marks the mobility of the catalytic subunit of this protein which can be identified by phosphorylation with radioactive ATP under appropriate conditions (Grefrath and Reynolds, 1973).

Clearly, it is essential to follow 'activity' of the intrinsic membrane protein of interest throughout all manipulations including the removal of extrinsic proteins. Care must also be exercised in classifying extrinsic proteins by the criterion of solubility in aqueous medium. It is possible to irreversibly denature and aggregate some extrinsic proteins such as actin during the process of removal from the bilayer. If the subsequent step is simply pelletting all 'non-soluble' material in a single centrifugation step, the aggregated protein may well appear with the membrane fraction, but not in fact be associated with it. A far preferable procedure is to separate the membrane fraction from aggregated released proteins by suitable density gradient centrifugation.

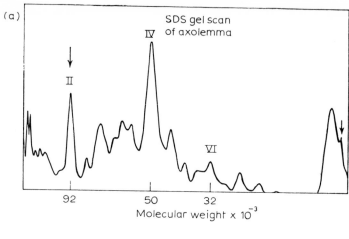

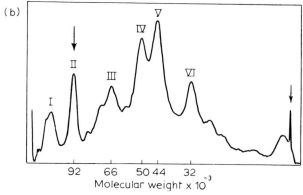

Fig. 2.3 (a) Densitometry scan of SDS gel of garfish axolemma stained with Coomassie Brilliant Blue. The membrane was purified in the presence of 0.1 M sodium bicarbonate, pH 7.9. (b) As (a) but the membrane was washed extensively with 0.001 M EDTA.

(b) *Interaction of detergents with membranes*

Addition of detergents (*D*) to an intact membrane containing intrinsic protein (*P*) and lipid (*L*) can be conveniently described by the following equilibria. These equations are greatly oversimplified but suffice to demonstrate the general types of interactions encountered.

$$(P-L)_{\text{bilayer}} + D \rightleftharpoons (P-L-D)_{\text{bilayer}} + D \tag{2.2}$$

$$(P-L-D)_{\text{bilayer}} + D \rightleftharpoons (P-L-D)_{\text{mixed micelle or bilayer fragment}}$$

$$+ (D-L)_{\text{mixed micelle}} + D \tag{2.3}$$

$$(P{-}L{-}D)_{\text{mixed micelle or membrane fragment}} + D \rightleftharpoons (P{-}D)_{\text{mixed micelle}}$$

$$+ (D{-}L)_{\text{mixed micelle}} + D \qquad (2.4)$$

$$mD \rightleftharpoons D_{\text{m}} \qquad (2.5)$$

Equation (2.2) is the partitioning of detergent into the lipid–protein bilayer and the final structure of the complex is governed by the large excess of lipid over detergent.

Equation (2.3) represents intermediate states in which the protein–lipid–detergent complex is either a large mixed micelle or a fragmented bilayer.

Equation (2.4) is the end state where a large excess of free detergent is present and all membrane protein and lipid exist in mixed micellar form with the detergent.

Equation (2.5) is presented to remind the reader that free detergent is in equilibrium with its own pure micelles at all concentrations of free D above the CMC. A similar equation should be written for all lipid species, but the free lipid monomer concentration is less than 10^{-10} M, much lower than the CMCs of detergents.

If membranes containing both extrinsic and intrinsic proteins are titrated with detergent, the extrinsic proteins are usually released in the first step (equation 2.2).

An elegant experimental example of the above equilibria is provided by the studies of Helenius and Simon (1975) on the interaction of detergents with Semliki Forest Virus.

A number of important points need to be emphasized with respect to the above equilibria.

(1) All the reactions are reversible.

(2) *Free* or unbound detergent is present in all steps, so that removal of D subsequent to solubilization will lead to reformation of the protein–lipid bilayer.

(3) In order to reproduce a solubilization experiment the concentrations of both membrane and total D added must be known. Altering either concentration will change the ratio of products unless the ratio of detergent to membrane is extremely high such that a large excess of detergent micelles is present.

Unfortunately, we cannot predict the equilibrium constants for the reactions listed. The partitioning of detergents into membranes and subsequent disruption to form mixed micelles is dependent on the chemical structure of the detergent and on the type and amount of protein and lipid in the membrane. Even if experimental data were available on the partitioning of detergents into pure lipid systems and the subsequent transition from bilayer to micellar structure, we would not know *a priori* the contribution of specific membrane proteins to the detergent partitioning.

Disruption of the intact bilayer described by equation (2.2)–(2.4) occurs by insertion of a single tail amphiphile and consequent disruption of the lipid head group packing and the order at the glycerol interface. At sufficiently high concentrations of detergent within the lipid bilayer the hydrocarbon interior becomes more accessible to water and reorganization must occur. The final mixed micelle of detergent in molar excess over lipid has a greater radius of curvature than the

Table 2.2 Statistical association

Detergent micelle: protein ratio	% 1 Copy*	% 2 Copies*
1:1	58.2	29.1
2:1	76.9	19.2
4:1	87.8	10.9
10:1	95.8	4.2

* Percentage of total protein present as 1 or 2 copies per micelle.

original bilayer structure and will vary in size depending upon the exact molar ratio of detergent to lipid.

If complete dispersion of membrane proteins in the state of one functioning unit per micelle is required, a greater than five- to tenfold excess of detergent micelles is required. For example, if only statistical considerations are invoked, Table 2.2 shows the number of copies of a single protein species which would be found per micelle as a function of micelle to protein ratio. Clearly, at low ratios one might find proteins apparently associated which in fact do not specifically interact with one another. Table 2.2 does not include lipid as a component of the system. Additional detergent is required to solubilize the lipid moieties of the membrane in micellar form. A reasonable estimate of total detergent needed to reach the equilibrium state of equation (2.4) in a given membrane system can be obtained by providing at least one micelle for each 5–10 lipid molecules together with the tenfold excess of micellar detergent required for total protein solubilization. Lower concentrations of total detergent will leave residual lipid bound to the proteins which in itself might be desirable if lipid is essential for maintenance of structure, but which also increases the probability of artifactual protein association.

Thus far, we have discussed only the thermodynamic aspects of detergent–lipid interactions. Another important aspect of such interactions involves the time required to reach equilibrium. Most studies of detergent solubilization do not include data which demonstrate attainment of equilibrium. Any one or more of the sequential solubilization steps may be extremely slow, especially at low temperatures, and failure of a particular detergent to solubilize totally may be due to kinetic considerations as well as to the nature of the detergent–membrane interaction. Lack of experimental data addressed to this point has led to much confusion and hampered attempts to systematize the information which is available on detergent–membrane interactions.

(c) *Titration of membranes with low concentrations of detergent*

Preliminary purification of some intrinsic membrane proteins can often be accomplished by the addition of detergents at concentrations which produce membrane fragments (equation 2.3). Jorgensen (1974) has developed this procedure for partial purification of a Na^+, K^+-ATPase from a number of membrane sources and found enrichment of the enzyme in specific membrane fractions separated by density gradient centrifugation.

Table 2.3 Density gradient fractionation of (Na$^+$, K$^+$) ATPase

Fraction	% Sucrose	Density (g cm^{-3})	Specific activity	% Total activity	% Total protein
0	9	1.0406	0	0	13
1	15	1.0620	4	1	5
2	25	1.1075	81	65	16
Pellet	>25	>1.1075	42	16	48

Membrane treated with SDS as shown in Fig. 2.4a was centrifuged on a discontinuous gradient at 78 000 g for 48 h at 4°C. The densitometer scan of an SDS gel of Fraction 2 is shown in Fig. 2.4b.

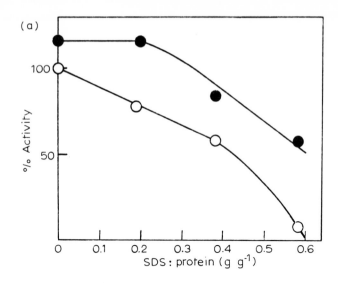

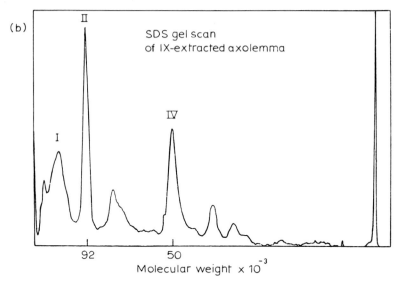

Fig. 2.4 (a) Titration of Na$^+$, K$^+$-ATPase from garfish olfactory axolemma with SDS in the presence ($\bullet$) and absence ($\circ$) of 2.5 x 10^{-3} M ATP. Protein concentration = 1.4 g l^{-1}; 0.002 M EDTA, 0.05 M imidazole HCl, pH 7.5. (b) Densitometry scan of SDS gel of garfish olfactory axolemma treated with SDS as shown in (a).

An example of the application of this technique is shown in Table 2.3 and Fig. 2.4a and b (Nielsen, 1977). Garfish olfactory axolemma which had been treated with

EDTA (Fig. 2.3a) was titrated with increasing concentrations of SDS and the total enzymatic activity of the Na^+, K^+-ATPase followed as shown in Fig. 2.4a. At 0.2 g SDS g^{-1} membrane protein (total concentration of membrane protein = 1.4 g l^{-1}) the enzyme retains full activity and is still in membrane bound form. Centrifugation of this mixture through a discontinuous sucrose density gradient results in fractionation of membrane fragments of differing density and protein composition as shown in Table 2.3. The SDS polyacrylamide gel scan of the fraction containing maximal amount of enzyme is shown in Fig. 2.4b and demonstrates significant purification of the enzyme as evidenced by the increase in relative amount of the catalytic subunit marked with an arrow. These membrane fragments contain bound SDS which is subsequently removed by washing in detergent-free buffer.

(d) *Solubilization of intrinsic membrane proteins in detergent micelles*
Addition of a large molar excess of detergent micelles to a membrane will lead to the final products in equation 2.4. In some cases endogenous lipid will remain with the protein until a very high molar ratio of detergent to membrane components is reached. For the purpose of purification and characterization there is no great virtue in removing all lipid, particularly since one cannot know *a priori* whether some specific portion of a lipid molecule is required for maintenance of native structure.

The question of which detergent to use is unanswerable given our present level of knowledge. It would seem reasonable to try to mimic the lipid bilayer as closely as possible which would mean using detergents containing *n*-alkyl chains of the approximate length of the fatty acyl chains in the natural lipid. This reasoning would suggest that bile salts and detergents with phenyl rings should be avoided. However, there are a number of examples showing that both of the latter types of detergent are able to maintain structure and function of some membrane-bound proteins (Karlin *et al.*, 1976; Reynolds and Stoeckenius, 1977; McCaslin and Tanford, 1979).

The head groups of available detergents listed in Table 2.1 are totally unlike those found in a native membrane with the exception of lysophospholipids. The large spatial extension of the non-ionic polyoxyethylenes gives some cause for concern in that this geometry might alter the water soluble domains of intrinsic membrane proteins through steric effects. However, there is ample experimental evidence that many membrane proteins retain their functional integrity in these detergent systems (Karlin *et al.*, 1976; Reynolds and Stoeckenius, 1977, Dean and Tanford, 1978). While there is at present no alternative to a trial and error approach with respect to which detergent is most suitable for a specific application, some miscellaneous information can be provided.

(1) Commercial detergents are heterogeneous mixtures of both hydrophobic tails and hydrophilic head groups. It is essential to know the composition of the compound chosen and actually used. Lubrol WX, for example, has a fraction with relatively short polyoxyethylene attached to C_{16} and C_{18} *n*-alkyl chains. This fraction tends to precipitate from concentrated aqueous solutions resulting in difficulty in reproducing chemical composition from one solution to another.

(2) Commercial detergents often contain impurities such as transition metal ions and auto-oxidation products of the detergent itself.

(3) All polyoxyethylene compounds are subject to auto-oxidation which should be prevented by the addition of small amounts of an anti-oxidant.

(4) Alkylphenyl hydrophobic groups absorb ultraviolet light and thus interfere with spectroscopic measurements.

(5) Ionic detergents with n-alkyl chains longer than 11 carbon atoms are, in general, denaturants of water-soluble proteins.

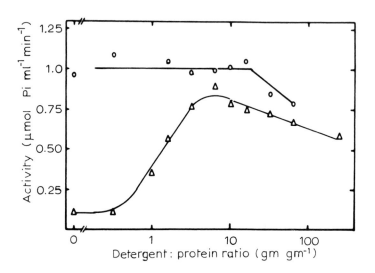

Fig. 2.5 Lubrol WX solubilization of Na$^+$ K$^+$-ATPase from shark rectal gland. Purified membrane bound enzyme was incubated for 30 min with increasing concentrations of lubrol WX at a constant protein concentration of 0.2 mg ml^{-1} in 3 M glycerol, 0.5 mM EDTA, 20 mM imidazole—HCl, pH 7.4. ○, Total activity in the mixture. △ Activity in the supernatant after centrifugation at 100 000 g for 1 h. (Reproduced with permission of D.F. Hastings, J.A. Reynolds and American Chemical Society.)

Despite our inability to generalize about specific detergent—membrane protein interactions, one can formulate some experimental rules. A systematic investigation of solubilization using a specific detergent and of the effect of this solubilization on function is absolutely essential. An example of the type of experimental data required is given here.

Fig. 2.5 shows the titration of rectal gland membrane from *Carcharhinus obscurus* with Lubrol WX (Hastings and Reynolds, 1979). The Na$^+$, K$^+$-ATPase activity in the total mixture is given together with the activity appearing in the supernatant after a 100 000 g centrifugation (1 h). This enzyme is maximally soluble

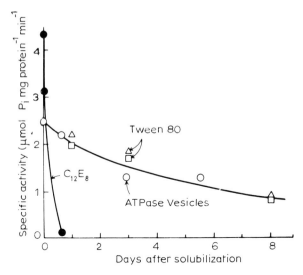

Fig. 2.6 Time dependence of Ca^{2+} ATPase activity measured at 23°C in
0.1 M KCl, 0.01 M Tes, pH 7.0, 10^{-4} M $CaCl_2$. All preparations were stored
in the cold between measurements. Open circles represent purified vesicles
without added detergent. Filled circles represent vesicles solubilized by
$C_{12} E_8$ at a detergent concentration of 10 mg ml^{-1}. Triangles and squares
represent two chromatographic fractions where $C_{12} E_8$ had been replaced
by Tween 80. (Reproduced with permission of Le Maire, Moller and Tanford,
and the American Chemical Society.)

and active at a detergent: protein ratio of 9:1 g/g (total protein concentration = 0.2 g l^{-1}).
At higher detergent: protein ratios a significant loss of activity is observed.

It is important that 'activity' measurements in systems containing detergent—
protein complexes be carried out without dilution. (Again refer to equations 2.2—2.4
which are reversible.) If an aliquot of solubilized protein in Fig. 2.5 is diluted into
an assay medium containing no detergent, protein—lipid bilayers will reform. If one
wishes to demonstrate that the protein—detergent complex is active in that state,
free and bound detergent must be present at the same levels as in the solubilization
experiments. This is most easily accomplished by not diluting the sample. If dilution
is experimentally essential, at the very least the assay medium should contain
detergent at the CMC to minimize shifts in the equilibrium state.

Once data of the type shown in Fig. 2.5 are obtained for a specific system, it is
necessary to determine the stability of the protein—detergent (—lipid) complex as a
function of time. Since further purification and characterization require days, it is
essential that the structure and function remain intact throughout that time period.
As has already been pointed out, kinetic considerations may result in metastable
states and the equilibrium state may not be attained during the time course of
solubilization.

One specific detergent may prove to be efficient as a membrane solubilizing agent, but the final protein–detergent (–lipid) complex obtained at equilibrium may be inactive. In this case it is often possible to solubilize in a metastable state in one detergent and quickly exchange for another detergent in which the complex is stable. An example of this is provided by studies with the Ca^{2+}-ATPase from sarcoplasmic reticulum. This membrane is rapidly solubilized by $C_{12}E_8$ but enzymatic activity is unstable in this detergent (Fig. 2.6). Le Maire *et al.* (1976) exchanged $C_{12}E_8$ for Tween 80 by gel chromatography and as shown in Fig. 2.6 the enzyme exhibits the same temporal stability in this form as in the intact membrane.

Exchange of detergents can also be carried out using density gradient centrifugation if the two detergents have sufficiently different partial specific volumes. Dialysis is not the method of choice in most cases since it is a slow procedure and micelles cannot difuse through the small pores required to retain most protein–detergent complexes.

In the discussion above it has been implicitly assumed that the detergents used are of a non-denaturing type. It has been known for many years that ionic detergents containing *n*-alkyl chains longer than 11 carbon atoms interact with water soluble proteins in a specific manner which leads to a conformational change and dissociation of oligomeric structures (Steinhardt and Reynolds, 1969). It is with some reluctance then that one considers the use of these detergents for membrane solubilization if retention of native structure is desired. We should keep in mind, however, the relationship between ligand binding and conformational states elucidated most clearly by Wyman (1964). If a protein has two accessible states which I shall call N and D, and both states can bind ligand, the equilibrium between the two states is governed by the following equation.

$$\frac{d \ln K}{d\, a_L} = \bar{v}_{D,L} - \bar{v}_{N,L} \qquad (2.6)$$

where K is the equilibrium constant for the reaction $N \rightleftharpoons D$, a_L is the activity of the ligand, $\bar{v}_{D,L}$ is moles ligand/mole D and $\bar{v}_{N,L}$ is moles ligand/mole N. If $\bar{v}_{N,L} \geqslant \bar{v}_{D,L}$ the protein will be stabilized in state N. Thus, if a membrane protein binds more detergent in a micellar form associated with its hydrophobic domain than in a denatured state, the membrane-bound native state will predominate. Cytochrome $b5$, for example, binds approximately one micelle of SDS at low detergent concentrations and is only denatured at very high detergent to protein ratios (Robinson and Tanford, 1975).

(e) *Purification of detergent–protein (–lipid) complexes*

A mixture of protein–detergent (–lipid) complexes in aqueous solution can be handled experimentally in the same manner as water soluble proteins with the proviso that free and bound detergent concentrations are not altered during any procedure.

The most powerful separation technique is affinity chromatography (Cuatrecasas and Anfinsen, 1971; Cuatrecasas, 1972). This procedure has been used to purify a large number of receptor proteins as well as a variety of glycoproteins. One specific type of affinity chromatography employs antibodies as the substrate co-valently attached to a solid support. At first glance this appears to be an attractive procedure for routine purification of a membrane protein which has already been obtained in small quantities by some other method. The principal disadvantage lies in the procedures required for elution of the bound protein–detergent (–lipid) complex from the matrix. Major alterations in pH or ionic strength are necessary to shift the binding equilibrium between antibody and protein, and these environmental changes may adversely affect the structure of the protein–detergent complex.

Ion exchange chromatography and gel filtration chromatography have also been used in purification of membrane proteins (Le Maire *et al.,* 1976; Hastings and Reynolds, 1979). Differential precipitation with salts such as ammonium sulfate is another common technique (Uesugi *et al.,* 1971), but is a somewhat risky procedure for detergent–protein complexes since one cannot predict the partitioning of detergent (and lipid) between the precipitated and aqueous phases.

One technique which has been neglected but holds considerable potential is that of free electrophoresis. Commercial equipment is available (designed primarily for concentrating water-soluble proteins) which is particularly convenient for the separation of complexes with differing charge to mass ratios (Allington *et al.,* 1978).

2.4 CHARACTERIZATION OF SOLUBILIZED MEMBRANE PROTEINS

The most fundamental criterion for retention of native structure in a protein–detergent (lipid) complex is retention of biological function and the response of that function to environmental variables. If the function is enzyme catalysis of a chemical reaction or binding of specific ligands, it is a straightforward procedure to compare these activities for soluble and membrane-bound forms of a protein. However, we are often concerned with proteins which are involved in transport or in communication from the extracellular to the intracellular space. In these systems activity can be determined only by reconstitution into a vesicular lipid system where a permeability barrier is present.

In view of the previous discussion of the primary difference between membrane and soluble forms of proteins with respect to an asymmetric environment, it is somewhat surprising how similar the functional properties are for those proteins which have been investigated (e.g. Kline *et al.,* 1971; Nielsen, 1977). Some of these studies, of course, have been carried out using 'leaky' vesicles in which part of the asymmetry of the membrane has been lost since the aqueous environment becomes identical on the inside and outside of freely-permeable vesicles.

Caution is required in interpretation of binding data as a criterion for retention of function. If the equilibrium between receptor and ligand is a one to one interaction, a direct comparison of membrane-bound with soluble protein is unambiguous. However, if the interaction is multivalent as in antibody—cell surface receptor systems, a direct comparison of binding data between cell—surface and solubilized protein is not possible quantitatively (Reynolds, 1979). The apparent affinity in the soluble system will be less than that in the membrane-bound system. This result is not a reflection of loss of binding activity but arises from the difference in bulk concentration of solubilized receptor and the surface concentration of membrane bound receptor.

Complete characterization of the solubilized membrane protein using chemical analyses is also essential, particularly when one wishes to determine the native molecular weight (Section 2.4.1).

(1) Quantitative amino acid analysis is probably the method of choice for obtaining total concentration of protein.

(2) Lipid concentration is determined by fatty acid analysis and by specific head group analysis.

(3) The types of lipid present in the final complex can be ascertained and is a potentially valuable piece of information. For example, one asks the question 'Does the protein retain a random distribution of endogenous lipid, or does it display specificity for particular lipids under the solubilization conditions employed?'.

(4) Analysis for bound detergent is tedious and requires large amounts of material unless one is fortunate enough to have radioactively-labelled amphiphiles. An excellent reference for analytical procedures is available (Rosen and Goldsmith, 1972).

(5) Carbohydrate determinations are most easily carried out using gas chromatography or high-pressure liquid chromatography.

Protein conformation in the soluble state can be investigated using a number of techniques. Chemical procedures have proven to be a valuable means of comparing membrane-bound and detergent-solubilized forms of a protein. The following examples are not exhaustive but serve to illustrate the application of these techniques.

(1) Limited proteolysis of the Ca^{2+}-ATPase from sarcoplasmic reticulum produces identical polypeptide fragments whether the enzyme is embedded in a lipid bilayer or dissolved in detergent micelles (Rizzolo and Tanford, 1978).

(2) An extensive investigation of susceptibility to proteolysis was carried out by Webster and co-workers on the coat protein from Fd bacteriophage in detergent micelles and in phospholipid vesicles (Chamberlain *et al.*, 1978). The results provide strong evidence for equivalent conformations in these two systems.

(3) An elegant study by Jorgensen has demonstrated that different proteolytic fragments are obtained from the Na^+ and K^+ forms of a $Na^+ K^+$-ATPase (Jorgensen, 1977). This investigation has not, however, been repeated with a soluble form of the enzyme.

Physical techniques can also be brought to bear on the problem of protein structure. One of the most attractive is circular dichroism since it requires very little material and is quite sensitive to alterations in protein structure. The use of this tool to deduce absolute values of secondary structure should be looked upon with some scepticism, but for comparative purposes it has great utility.

(1) Fd coat protein has identical circular dichroic spectra in phospholipid vesicles and in detergent micelles (Nozaki *et al.*, 1976). It has been inferred from these studies that this protein has a similar conformation in the host membrane of *E. coli*. On the other hand, the circular dichroism of this protein in the assembled virus differs significantly and appears to be that of a totally helical conformation. This data would suggest a major structural alteration as the coat protein proceeds from the bacterial membrane to the viral coat during morphogenesis.

(2) Bacteriorhodopsin has been shown to be approximately 70% helical in its native state (Henderson and Unwin, 1975). When solubilized in Triton X-100 this protein has a circular dichroic spectrum compatible with this high amount of helical structure and is monomeric (Reynolds and Stoeckenius, 1977). Circular dichroism in the visible wavelength region has been used to follow the loss of chromophore—chromophore interaction when bacteriorhodopsin is solubilized (Becher and Ebrey, 1976).

Fluorescence spectroscopy has been used extensively in investigations of intact membrane systems. However, it has thus far been of limited applicability in structural studies on solubilized protein—detergent (—lipid) complexes. A recent review by Waggoner (1976) discusses applications and problems in using extrinsic fluorescent probes but does not specifically deal with solubilized forms of intrinsic membrane proteins. Barrantes (1978) has used intrinsic fluorescence of acetylcholine receptor in its membrane environment to follow alterations in structure cause by agonist binding, and Wu and Stryer (1972) applied energy transfer techniques to the study of distances between fluorescent chromophores in rhodopsin.

Nuclear magnetic resonance which has been extremely valuable in investigations of pure lipid bilayers has not been used significantly in the study of protein conformation in protein—detergent (—lipid) complexes. The technical and theoretical problems in applying this technique to these complex systems can best be appreciated by reading recent reviews on the subject (e.g. Seelig, 1977).

Hydrodynamic measurements of particles in solution have long been used by the protein chemist to give some indication of shape. A combination of sedimentation velocity and sedimentation equilibrium provides a direct measurement of a hydrodynamic radius, R_s.

$$s = \frac{M_p \ (1-\phi'\rho)}{6\pi\eta R_s N} \tag{2.7}$$

where s is the sedimentation coefficient, $M_p \ (1-\phi'\rho)$ is measured directly from sedimentation equilibrium data, η is the solvent viscosity, and N is Avogadro's number. If the protein molecular weight and particle composition are known, a minimum radius for an unhydrated sphere can be calculated.

$$R_{min} = \left[\frac{3M \text{ (particle)} \bar{v} \text{ (particle)}}{4\pi N} \right]^{1/3} \tag{2.8}$$

A comparison of the measured R_s with R_{min} indicates a deviation from spherical shape and also a degree of hydration. (Note that these two variables are not separable unless the degree of hydration is determined by a separate experiment.) Most small globular proteins have R_s/R_{min} values of approximately 1.2. Small oblate ellipsoid micelles are slightly more asymmetric (and/or) hydrated with R_s/R_{min} ranging from 1.2 to 1.5. Measurements of R_s for membrane protein–detergent (–lipid) complexes have been reported but unambiguous interpretation of this data is not possible. It is easy to visualize a relatively asymmetric detergent micelle which is bound to an elongated membrane protein such that the total particle appears by hydrodynamic criterion to be symmetric.

Low angle X-ray scattering and neutron scattering have also been applied to the study of protein–detergent (–lipid) complexes. The principal advantage of these techniques is the ability to determine the relative centers of mass of the components (Sardet *et al.*, 1976; Oxborne *et al.*, 1978). The specialized equipment required for these experiments, however, is usually not available to the average researcher.

2.4.1 Molecular weight of solubilized native protein

Sedimentation equilibrium is the method of choice for the determination of molecular weight of the protein moiety in a protein–detergent (–lipid) complex. A photoelectric scanner is essential for this measurement since other optical systems will detect all components present in the solution rather than the protein moiety exclusively. Measurements can be made with as little as 0.1 ml of a solution of absorbance at an appropriate wavelength as low as 0.05.

At equilibrium the concentration of protein in the complex particle is measured as a function of the distance from the center of axial rotation.

$$\frac{d \ln c}{dr^2} = M_p (1 - \phi' \rho) = M_p [(1 - \bar{v}_p \rho) + \sum_i \delta_i (1 - \bar{v}_i \rho)] \tag{2.9}$$

where c = concentration of protein, r = distance from the center of rotation, M_p = molecular weight of protein, $\bar{v}_p$ = partial specific volume of the protein, δ_i = g component i/g protein, $\bar{v}_i$ = partial specific volume of component i, and ρ = solvent density. Chemical analysis provides δ_i for lipid if present and for detergent. $\bar{v}$ has been tabulated for a large number of detergents (Steele *et al.*, 1978) and $\bar{v}_p$ is obtained with sufficient accuracy from the amino acid composition.

Table 2.4 Molecular weights

Protein	Detergent	Native M_p	Polypeptide M_p	Polypeptide particles	Reference
Glycophorin	$C_{12}OSO_3^-$	14 000	14 000	1	Grefrath and Reynolds (1974)
Cytochrome b5	DOC^-	16 200	16 200	1	
	Triton X-100	16 200	16 200	1	Robinson and Tanford (1965)
Ca^{2+}-ATPase	Tween 80	400 000	119 000	4	Le Maire et al. (1976); Rizzolo et al. (1976)
	$C_{12}E_8$	119 000–238 000	119 000	1–2	Dean and Tanford (1978)
Bacteriorhodopsin	Triton X-100	24 000	24 000	1	Reynolds and Stoeckenius (1977)
Acetylcholine receptor	Brij 58	250 000–500 000			
$Na^+ K^+$-ATPase	Lubrol WX	360 000	106 400	2	Hastings and Reynolds (1979)
			36 600	4	
Fd coat protein	$C_{12}OSO_3^-$	10 300	5 200	2	Makino et al. (1975)
Rhodospin	Cholate	117 000–156 000	39 000	3–4	McCaslin and Tanford (1979)
Thy-1	Brij 96	12 500	12 500	1	

If the detergent concentration bound to the protein is unknown, it is still possible to obtain M_p (Reynolds and Tanford, 1976) by determining $M_p (1-\phi'\rho)$ as a function of solvent density using $D_2O:H_2O$ mixtures. At $\rho = 1/\bar{v}_d$ the buoyant density term due to the detergent is zero and M_p is obtained directly from equation (2.9).

Since the binding of detergent to the membrane protein is a reversible process, it is important that the centrifuge speed be selected such that free detergent micelles do not sediment appreciably.

The molecular weights of a number of functioning membrane proteins solubilized in detergent have been determined as described above and are listed in Table 2.4. Only those proteins which have prosthetic groups absorbing in the visible wavelength range can be studied in detergents containing phenyl rings such as Triton X-100. Furthermore, the density blanking procedure cannot be used with most of the bile salts since they have $\bar{v}$ values close to those of proteins.

It must be emphasized that gel filtration chromatography and polyacrylamide gel electrophoresis do not measure molecular weights and that sucrose density gradient centrifugation cannot be used since the solvent composition then differs from the composition of bound water and the buoyant density term due to bound water is not equal to zero.

2.4.2 Molecular weight of constituent polypeptides

The molecular weight of individual polypeptides which constitute the native protein are determined as described in the previous section. It is necessary to dissociate these polypeptides from one another and to separate non-identical species. This is usually accomplished in denaturing detergents such as SDS keeping in mind that not all oligomeric membrane proteins are dissociated by this detergent. Separation of unlike polypeptide chains is carried out by gel filtration chromatography, gel electrophoresis, or some type of affinity chromatography. SDS binding can be measured using radioactive detergent. Table 2.4 gives the molecular weights of constituent polypeptides of a number of intrinsic membrane proteins. In some cases the sequence has been reported and agrees well with the molecular weight determined in the presence of SDS by analytical ultracentrifugation.

Guanidine hydrochloride has been used successfully with water soluble proteins as a denaturing and dissociating agent. However, this solvent has proven to be a poor choice for intrinsic membrane proteins which often retain ordered structure and are self-associated at concentrations of the denaturant as high as 7.0 M.

2.5 CONCLUSIONS

In this chapter I have attempted to summarize some general principles of detergent solubilization of membrane proteins and the characterization of these complexes. As was stated previously, we are in no position at this time to provide absolute guidelines as to the choice of detergent for a particular system. However, the methods of rational and systematic procedure are clear-cut and one hopes that in the near future sufficient experimental data will become available that we will understand what features of particular detergents allow retention of native structure in membrane–protein–detergent complexes.

The importance of demonstrating that one has indeed maintained native function in soluble form cannot be over-emphasized. Characterization of an altered, non-native protein state is obviously not particularly useful when we wish to understand the *in vivo* system. In defining a native, soluble state it is important to consider the total system being investigated, namely, an aqueous solution containing excess detergent micelles and a detergent–protein (–lipid) complex. A substrate, for example, may partition into detergent micelles thus altering the concentration in the aqueous solution which is 'accessible' to the protein being studied. Detergents may bind small inorganic cations which are required for protein 'activity' thus providing competitive binding sites for these chemical substances.

The demonstration of native function also has implications in reconstitution experiments. Reconstitution is accomplished by reversal of the equilibrium equations given in Section 2.3.2(b), and requires removal of the detergent from the solubilized system. There has been an unfortunate tendency to reconstitute from detergent systems known to denature the protein being studied simply because these particular detergents have high CMCs and are easily dialyzed (e.g. cholate and deoxycholate). Such a procedure leaves open to serious question the results of such experiments. It is apparent that a great deal of experimental work remains to be done aimed at devising appropriate reconstitution methods from non-denaturing detergents which usually have the unfortunate property of possessing low CMCs (e.g. polyoxyethylene derivatives). Gel filtration chromatography and density gradient centrifugation in the absence of free detergent are applicable to these systems, but little experimental data are available in the literature relating to these methods.

Finally, I reiterate the importance of not confusing thermodynamic and kinetic data. Statements are often made that a particular detergent does not solubilize a specific membrane protein. Does this mean that the final equilibrium state is a lipid bilayer containing protein and dissolved detergent molecules, or is one, in fact, dealing with a slow kinetic rate such that the equilibrium state was not reached in the time course of the experiment?

REFERENCES

Allington, W.B., Cordry, A.L., McCullough, G.A., Mitchell, D.E. and Nelson, J.W. (1978), *Anal. Biochem.*, **85**, 188--196.
Barrantes, F.J. (1978), *J. mol. Biol.*, **124**, 1–26.
Becher, B. and Ebrey, T.G. (1976), *Biochem. biophys. Res. Commun.*, **69**, 1–6.
Bretscher, M.S. (1973), *Science*, **181**, 622–629.

Chamberlain, B.K., Nozaki, Y., Tanford, C. and Webster, R.E. (1978), *Biochim. biophys. Acta,* **510**, 18−37.

Cuatrecasas, P. (1972), *Advan. Enzymol.,* **36**, 29.

Cuatrecasas, P. and Anfinsen, C.B. (1971), *Ann. Rev. Biochem.,* **40**, 259−278.

Cuatrecasas, P. and Hollenberg, M.D. (1976), *Adv. prot. Chem.,* **30**, 252−451.

Dean, W.L. and Tanford, C. (1978), *Biochemistry,* **17**, 1683−1690.

Green, D.E. (1972), *Ann. N.Y. Acad. Sci.,* **195**, 150−173.

Grefrath, S.P. and Reynolds, J.A. (1973), *J. biol. Chem.,* **248**, 6091−6094.

Grefrath, S.P. and Reynolds, J.A. (1974), *Proc. natn. Acad. Sci. U.S.A.,* **71**, 3913−3916.

Hastings, D.F. and Reynolds, J.A. (1979), *Biochemistry,* **18**, 817−821.

Helenius, A., McCaslin, D.R., Fries, E. and Tanford, C. (1979), *Meth. Enz.,* **56**, 734−749.

Helenius, A. and Simons, K. (1975), *Biochim. biophys. Acta,* **415**, 29−79.

Henderson, R. and Unwin, P.N.T. (1975), *Nature,* **257**, 28−31.

Jorgensen, P.L. (1974), *Biochim. biophys. Acta,* **356**, 36−52.

Jorgensen, P.L. (1977), *Biochim. biophys. Acta,* **466**, 97−108.

Karlin, A., McNamee, M.G., Weill, C.L. and Valderrama, R. (1976), In: *Methods in Receptor Research* (Blecher, M. ed.), Marcel Dekker, New York.

Kline, M.H., Hexum, T.D., Dahl, J.L. and Hokin, L.E. (1971), *Arch. biochem. Biophys.,* **147**, 781−787.

Korn, E.D. (1978), *Proc. natn. Acad. Sci., U.S.A.,* **75**, 588−599.

Kuchel, P.W., Campbell, D.G., Barclay, A.N. and Williams, A.F. (1978), *Biochem J.*

Le Maire, M., Moller, J.V. and Tanford, C. (1976) *Biochemistry,* **15**, 2336−2342.

McCaslin, D.R. and Tanford, C. (1979), *Biochemistry,* In press.

Makino, S., Woolford, J.L., Jr., Tanford, C. and Webster, R.E. (1975), *J. biol. Chem.,* **250**, 4327−4332.

Marchesi, V.T. and Palade, G.E. (1967), *J. cell Biol.,* **35**, 385.

Nielsen, T.B. (1977), *Ph. D. Dissertation,* Duke University, Durham, North Carolina.

Nozaki, Y., Chamberlain, B.K., Webster, R.E. and Tanford, C. (1976), *Nature,* **259**, 335−337.

Oxborne, H.B., Sardet, C., Michel-Villaz, M. and Chabre, M. (1978), *J. mol. Biol.,* **123**, 177−206.

Reynolds, J.A. (1979), *Biochemistry,* **18**, 264−269.

Reynolds, J.A. and Karlin, A. (1978), *Biochemistry,* **17**, 2035−2038.

Reynolds, J.A. and Stoeckenius, W. (1977), *Proc. natn. Acad. Sci., U.S.A.,* **74**, 2803−2804.

Reynolds, J.A. and Tanford, C. (1976), *Proc. natn. Acad. Sci., U.S.A.,* **73**, 4467−4470.

Reynolds, J.A. and Trayer, H. (1971), *J. biol. Chem.,* **246**, 7337−7342.

Rizzolo, L.J., Le Maire, M., Reynolds, J.A. and Tanford, C. (1976), *Biochemistry,* **15**, 3433−3437.

Rizzolo, L.J. and Tanford, C. (1978), *Biochemistry,* **17**, 4049−4055.

Robinson, N.C. and Tanford, C. (1975), *Biochemistry,* **14**, 369−378.

Rosen, M.J. and Goldsmith, H.A. (1972), *Systematic Analysis of Surface Active Agents,* John Wiley and Sons, New York.

Sardet, C., TardienA. and Luzzati, V. (1976), *J. mol. Biol.,* **105**, 383–407.

Seelig, J. (1977), *Quart. Rev. Biophys.,* **10**, 353–418.

Singer, S.J. and Nicolson, G.L. (1972), *Science,* **175**, 720–731.

Small, D.M. (1971), In: *The Bile Acids* (Naire, P.P. and Kritchevsky, D., eds.), Vol. 1, 249–356, Plenum Press, New York.

Steele, J.C.H., Tanford, C. and Reynolds, J.A. (1978), *Meth. Enz.,* **48**, 11–23.

Steinhardt, J. and Reynolds, J.A. (1969), *Multiple Equilibria in Proteins,* Academic Press, New York.

Tanford, C. (1973), *The Hydrophobic Effect,* John Wiley and Sons, New York.

Tanford, C. and Reynolds, J.A. (1976), *Biochim. biophys. Acta,* **457**, 133–170.

Tanford, C. (1974), *J. phys. Chem.,* **78**, 2469–2479.

Uesugi, S., Dulak, N.C., Dixon, J.F., Hexum, T.D., Dahl, J.L., Perdue, J.F. and Hokin, L.E. (1971), *J. biol. Chem.,* **246**, 531–543.

Waggoner, A. (1976), In: *The Enzymes of Biological Membranes,* Vol. 1, 119–137, Plenum Press, New York.

Wu, C.W. and Stryer, L. (1972), *Proc. natn. Acad. Sci. U.S.A.,* **69**, 1104–1108.

Wyman, J. (1964), *Advan. prot. Chem.,* **19**, 224–286.

3 Affinity Chromatography for Membrane Receptor Purification

STEVEN JACOBS and PEDRO CUATRECASAS

Membrane Receptors : Methods for Purification and Characterization
(*Receptors and Recognition,* Series B, Volume 11)
Edited by S. Jacobs and P. Cuatrecasas
Published in 1981 by Chapman and Hall, 11 New Fetter Lane, London EC4P 4EE
© Chapman and Hall

3.1 INTRODUCTION

Membrane receptors possess no unique gross physical or chemical properties that distinguish them from other membrane proteins of which they constitute only a small fraction of a percent. Their most remarkable characteristic is their ability to bind specific ligands with high affinity and to translate this binding into a relevant biological response. If not for this property, receptors would remain obscure in the sea of the other more numerous proteins that surround them. Therefore, it is not surprising that conventional methods of protein purification, which separate molecules on the basis of physical–chemical properties such as size, shape, charge and solubility, have only limited usefulness for purifying receptors, while affinity chromatography (Cuatrecasas, 1970; Cuatrecasas and Anfinsen, 1971), which depends upon biospecific recognition, is an extremely powerful tool. Table 3.1 lists some of the membrane receptors that have been purified using affinity chromatography.

The basic principle of affinity chromatography is illustrated in Fig. 3.1. In its simplest form, a ligand (e.g. a hormone) that specifically binds with high affinity to the receptor to be purified is covalently attached to a solid support. A crude mixture containing the solubilized receptor is passed over the affinity support. The receptor is tightly adsorbed to the support as a result of binding to the ligand. Impurities are removed by washing. The receptor is then eluted using various techniques.

3.2 AFFINITY ADSORBENTS

An affinity adsorbent has three basic components: an insoluble matrix, a ligand that will bind to the receptor, and a spacer arm. In addition, a suitable method of coupling these components is required.

3.2.1 The matrix

The matrix must be an insoluble material with chemically reactive groups for activation and coupling. It must be mechanically and chemically stable under conditions of activation and coupling and during operation. Such conditions may entail organic solvents, denaturants such as urea or guanidine, extremes of pH, oxidizing or reducing agents, or the presence of degradative enzymes. In addition, if the adsorbent is to be stored for any length of time, resistance to microbial attack is desirable. Ideally, the matrix must provide a large adsorbtive surface area for which a high degree of permeability is useful. If the adsorbent is to be used in a column rather than in a batch procedure, good flow properties are important. For this

Table 3.1

Receptor	Immobilized ligand	Method of elution	Reference
Nicotinic acetylcholine	Quaternery amonium	NaCl Gradient, Bis-Q, Carbamylcholine, or flaxedil	Schmidt and Raftery, 1972; Chang, 1974; Karlin and Cowburn, 1973; Olsen et al., 1972
Nicotinic acetylcholine	Naja naja toxin	Carbamylcholine	Onj and Brady, 1974; Karlsson et al., 1972
Nicotinic acetylcholine	Cobratoxin	Carbamylcholine	Brockers and Hall, 1975
Nicotinic	Erabutoxin-B	Decamethonium	Merlie et al., 1975
Insulin	Insulin	Urea, pH 6.0	Cuatrecasas, 1973; Jacobs et al., 1977
Insulin	Concanavalin A	α-Methyl mannopyranoside	Jacobs et al., 1977; Cuatrecasas and Tell, 1973
Insulin	Wheat germ agglutinin	N-Acetyl glucosamine	Cuatrecasas and Tell, 1973
hCG/LH	hCG	0.025 M Acetic acid	Dufau et al., 1975
Prolactin	hGH	5 M $MgCl_2$	Shiu and Friesen, 1975
Growth hormone	hGH	5 M $MgCl_2$	McIntosh et al., 1976
TSH	TSH	0.2 M Acetate, pH 2.5	Tate et al., 1975
β-Adrenergic	Alprenalol	Alprenalol and 1M NaCl or alprenalol	Vanquelin et al., 1977; Caron et al., 1979
EGF	Anti EGF antibody	10% Formic acid	Hock et al., 1979

(continued on the next page)

Table 3.1 (continued)

Receptor	Immobilized ligand	Method of elution	Reference
EGF	Wheat germ agglutinin and other lectins	N-Acetyl glucosamine or the specific sugar	Hock et al., 1979
Asialoglycoprotein	Asialo-orosomucoid	Calcium free 1.25 M NaCl	Hudgin et al., 1974
Mannan-binding protein	Mannan	EDTA	Kawasaki et al., 1978
IgE Receptor	IgE	KSCN	Kawasaki et al., 1978
Intrinsic factor	B12–I.F. complex	EDTA or pH 5.0	Marcoullis and Grasbeck, 1977; Kouvonen and Grasbeck, 1979
Transcobalamin II	B12–Transcobalamin II	EDTA, pH 5.0	Seligman and Allen, 1978
Platelet von Willibrand factor	Wheat germ agglutinin	N-Acetylglucosamine	Cooper et al., 1979

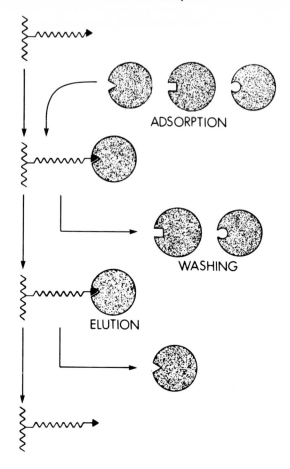

ADSORPTION

WASHING

ELUTION

Fig. 3.1 Principle of affinity chromatography.

purpose rigid spherical beads are the best physical form. It is particularly important
that the matrix be free of ionic or hydrophobic sites that could act as non-specific
adsorption centers.

Although no known material possesses all these properties, several matrices are
in common use and have proven suitable. Cellulose, which was one of the earliest
matrices used, has several undesirable properties. Its fibrous structure results in poor
flow and restricts the permeability of large proteins. Furthermore, cellulose often
exhibits considerable non-specific adsorption, particularly after the introduction
of charged groups during activation. Porous glass has the advantages of being
resistant to acids and organic solvents. Because of its rigidity it has excellent flow
properties, but non-specific adsorption is often considerable due to SiOH groups,
limiting its usefulness. Polyacrylamide beads are composed of a covalently crosslinked

network that gives them a high degree of physical and chemical stability. The most outstanding property of polyacrylamide is the large number of chemically reactive carboxamide groups which allows for a higher degree of substitution than is available with other common matrices (Inman and Dintzis, 1969; Inman, 1974). Unfortunately, shrinkage and decrease in permeability occur during derivatization of polyacrylamide, making it impermeable to large macromolecules (Cuatrecasas, 1970) such as receptors. Thus it may be possible to derivatize polyacrylamide with very high concentrations of ligand, much of which is inaccessible to the receptor being purified. This can greatly exacerbate the problem of leakage. Crosslinked dextrans that are sufficiently permeable for receptors are too soft for column procedures.

For most purposes, agrose or certain derivatives of agarose appear to be the most suitable affinity matrices that are in common use. Agarose is a polymer of D-galactopyranosyl and 3, 6-anhydro-L-galactopyranosyl units substituted to a minor degree with sulfate esters. It has a tertiary structure of double helices alternating with single chains in random coil conformation. This structure gives it rigidity and also extreme porosity which makes it permeable to large molecules (Rees, 1969). It is comparatively free of non-specific adsorption centers. Several mild procedures have been devised for coupling proteins or small ligands (Cuatrecasas, 1970; Cuatrecasas and Anfinsen, 1971; Parikh *et al.*, 1974). Nevertheless, agarose has some unfavorable properties. At temperatures above 40°C the gel beads dissolve. Even at room temperature some leakage of soluble carbohydrate occurs. This is accelerated below pH 2.5 and by various organic solvents and urea (Porath and Kristiansen, 1975).

By crosslinking agarose with reagents such as epihalohydrins (Porath *et al.*, 1971), bisoxyranes (Porath *et al.*, 1971), or divinylsulfones (Porath *et al.*, 1975), many of these undesirable properties can be eliminated. For example, epichlorhydrin-crosslinked beads are stable to strong alkali, organic solvents, and chaotropic agents (Porath *et al.*, 1971). They can be boiled or autoclaved (Porath *et al.*, 1971). They retain their shape after lyophilization and rehydration (Porath *et al.*, 1971). Because they are more rigid than the parent agarose bead, the crosslinked derivative can be subjected to higher hydrostatic pressure without compression, allowing higher operating flow rates. Furthermore, by treating crosslinked agarose with strong alkali under reducing conditions, the sulfate esters present in natural agarose are cleaved, decreasing their already low non-specific ionic-adsorption properties (Porath *et al.*, 1971). Crosslinked, desulfated agarose beads are commercially available (Sepharose CL).

3.2.2 Ligands

A ligand which is to be immobilized for use in affinity chromatography must have a substantial affinity for the receptor that is to be purified and also a chemically reactive group through which it can be coupled to the matrix without severe loss of affinity. Ideally, the ligand should be highly specific for the receptor; however, ligands of relatively low specificity may be adequate if a highly specific method of elution can be devised.

(a) *Agonists and competitive antagonists*
Various categories of ligands can be immobilized for use in affinity chromatography. The endogenous substance that the receptor naturally recognizes (e.g., hormone or neurotransmitter) is an obvious candidate and has been used frequently and success-fully (Table 3.1). Sometimes a competitive agonist or antagonist has even greater specificity, more favorable affinity, or may be more easily immobilized. This seems to be particularly true for small molecules. For example, the catechol moiety of naturally occurring β-adrenergic catacholamines is recognized by a membrane binding site other than the β-adrenergic receptor (Cuatrecasas *et al.*, 1974). Alprenalol, a synthetic competitive antagonist that is more specific and has higher affinity for the β-adrenergic receptor, has been used in its purification (Vanquelin *et al.*, 1977; Caron *et al.*, 1979). Also, neurotoxins and quaternary ammonium compounds which compete with acetylcholine have been used to purify nicotinic receptors (Table 3.1).

(b) *Lectins*
Since most cell surface receptors appear to be glycoproteins, lectins, which are naturally occurring carbohydrate-binding proteins, may have general applicability as affinity ligands in the purification of these receptors. Concanavalin A and wheat germ agglutinin affinity columns have been used to purify insulin receptors (Jacobs *et al.*, 1977; Cuatrecasas and Tell, 1973). TSH receptors that have been solubilized with lithium iodosalicylate adsorb to Concanavalin A-agarose (Tate *et al.*, 1975). One major advantage of lectin-affinity columns is that the adsorbed receptor can be eluted by using a simple sugar. This is a mild biospecific procedure which is unlikely to result in denaturation of the receptor or to interfere with its subsequent assay. In addition, lectins bind to many receptors at a site distinct from that to which the endogenous, naturally occurring ligand binds. Furthermore, lectin binding does not stringently depend upon the tertiary structure of the receptor protein. This may be used to advantage. For example, the binding activity of the receptor for epidermal growth factor is lost when it is solubilized with non-ionic detergents. However, covalent labeling of the receptor with an arylazide derivative of ^{125}I-EGF followed by solubilization does not interfere with the subsequent purification of the receptor using lectin affinity columns (Hock *et al.*, 1979).

Since the receptor to be purified probably comprises only a small fraction of the total membrane protein to which the lectin binds, it might be anticipated that a lectin column would lack sufficient specificity to be useful. In practice, this is frequently not the case. For example, although insulin receptors comprise only about 0.1% of the total Concanavalin A binding sites of liver membranes, a 200 000-fold purification has been obtained using Concanavalin A-agarose (Cuatrecasas and Tell, 1973).

Several lectins and their specific sugars are listed by Kristiansen (1974).

(c) *Antibodies*

Immuno-adsorbents have been used to purify a large number of proteins and polypeptides and have potential use in the purification of membrane receptors. Antibodies directed against surface membrane receptors occur spontaneously in some patients with auto immune disorders. Autoantibodies directed against nicotinic receptors (Almon *et al.*, 1974; Lindstrom, 1976; Aharonov, 1975), insulin receptors (Flier, 1975) and thyroid-stimulating hormone receptors (Smith and Hall, 1974) have been described, and it is likely that autoantibodies directed against other receptors may be discovered in the future. Also, antibodies directed against membrane receptors have been produced by immunizing with solubilized receptor preparations of varying degrees of purity. These include insulin receptors (Jacobs *et al.*, 1978), nicotinic receptors (Patrick *et al.*, 1973; Valderama *et al.*, 1976; Aharonov *et al.*, 1977; Lindstrom *et al.*, 1977), prolactin receptors (Shiu and Friesen, 1976), and growth hormone receptors (Waters and Friesen, 1978). The specificity of most of these anti-receptor antisera have not been described. Therefore, their usefulness for affinity chromatography is not clear. In addition, the conditions frequently required to elute proteins from antibody columns may irreversibly denature most receptors. The development of monoclonal antibodies to membrane receptors is likely to increase the usefulness of immunoadsorbents for receptor purification.

A novel approach to the use of antibodies for membrane receptor purification has recently been described (Hock *et al.*, 1979). A photoaffinity-labeled derivative of epidermal growth factor was used to covalently couple epidermal growth factor to its receptors. Antibodies to epidermal growth factor were immobilized on agarose and used to adsorb the epidermal growth factor–receptor complex. This method, which may have general applicability, will be discussed in more detail later in this chapter.

(d) *Hydrophobic chromatography*

Integral membrane proteins have hydrophobic domains that are normally embedded in the lipid bilayer. Through these domains, they will frequently adsorb to agarose that has been derivatized with hydrophobic groups. The use of such derivatives is called hydrophobic chromatography (Er-el *et al.*, 1972; Shaltiel, 1974).

In one of the earliest examples of hydrophobic chromatography, a homologous series of hydrocarbon derivatives of Sepharose (Seph-NH$(CH_2)_n$H), which differed only in the length of the alkyl side chain, was prepared (Er-el *et al.*, 1972). It was found that glycogen phosphorylase was not retained on Seph-Cl, was retarded on Seph-C2 and Seph-C3, and was tightly adsorbed on Seph C4–6. This kind of pattern is seen repeatedly in hydrophobic chromatography. That is, the longer the hydro-phobic group, the more tightly the protein is bound. By testing a homologous series of immobilized hydrophobic derivatives, a derivative can be chosen whose length results in adsorption of the protein of interest and yet permits elution under sufficiently mild conditions.

In some cases, the mechanism of 'Hydrophic chromatography', in addition to simple hydrophobic binding, depends upon an interaction with charged groups that are either purposely introduced into the hydrophobic derivative or adventitiously result from the method of coupling (Hofstee, 1973; Yon, 1977; Jost *et al.*, 1974). The combined hydrophobic and polar interaction is analogous to the interaction of the protein with phospholipids or detergents. In fact, detergents have been immobilized and used to fractionate membrane proteins (Cresswell, 1979).

For reasons that will be discussed in Section 3.2.4, the degree of substitution of the matrix is of critical importance in hydrophobic chromatography. Therefore, the degree of substitution should be carefully controlled (Homcy *et al.*, 1977).

Charge-free hydrophobic derivatives of agarose have been used to purify adenylate cyclase (Homcy *et al.*, 1977). HLA antigens have been purified using immobilized deoxycholate (Cresswell, 1979).

3.2.3 Spacer arms

Frequently the avidity of an affinity adsorbent will be increased if the ligand is attached to the insoluble matrix via a flexible spacer arm, rather than directly. The spacer arm, which appears to be particularly important when a small ligand is coupled for the purification of a high molecular weight protein, is required to overcome steric hindrance by the matrix. Several examples of this phenomenon are known. For example, Sepharose-bound D-tryptophan methyl ester is far less effective than the Sepharose-ϵ-aminocaproyl-D-tryptophan methyl ester in the purification of α-chymotrypsin (Cuatrecasas *et al.*, 1968). The capacity of Sepharose for staphylococal nuclease is increased 4- to 5-fold if the competitive inhibitor, pUTp-aminophenyl, is attached to Sepharose via a 10 Å spacer rather than attaching it directly (Cuatrecasas, 1970). The requirement for a spacer arm is even clearer in the case of p-aminophenyl-β-D-thioglactopyranoside when used in the purification of β-galactosidase. Sepharose derivatives with this inhibitor coupled directly fail to bind β-galactosidase. However, derivatives with this inhibitor coupled via a 20 Å spacer retained the enzyme tightly (Steers *et al.*, 1971). At least part of this effect has been shown to be due to a non-specific interaction of β-galactosidase with the hydrophobic spacer rather than simply overcoming steric hindrance (O'Cara *et al.*, 1974). Insulin agarose is far more effective in binding insulin receptors when coupled via a spacer arm (Cuatrecasas, 1973).

For a particular application the need for a spacer arm, as well as the optimal length and chemical nature of the arm, must be determined empirically. Spacer arms often introduce non-specific hydrophobic or ionic adsorption centers. Therefore, it is usually best to use a hydrophilic, uncharged spacer arm which is not longer than is necessary to overcome steric hindrance. Although rare, another complication may occur if excessively long spacer arms are used: the arm may fold over on itself, burying the ligand in the matrix. This may explain why, in some systems (Lowe *et al.*, 1973), increasing the length of the spacer up to a certain length will increase the binding strength of the adsorbent, but further increases in length will abruptly cause a decrease in binding strength.

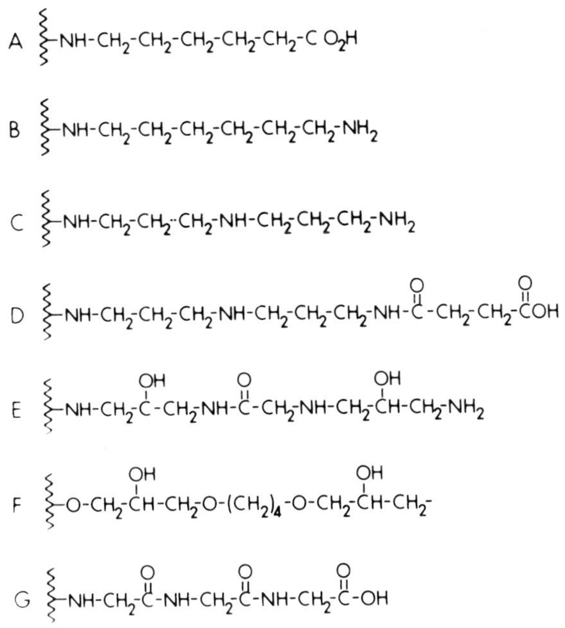

A $-NH-CH_2-CH_2-CH_2-CH_2-CH_2-C\,O_2H$

B $-NH-CH_2-CH_2-CH_2-CH_2-CH_2-CH_2-NH_2$

C $-NH-CH_2-CH_2-CH_2-NH-CH_2-CH_2-CH_2-NH_2$

D $-NH-CH_2-CH_2-CH_2-NH-CH_2-CH_2-CH_2-NH-\overset{O}{\overset{\|}{C}}-CH_2-CH_2-\overset{O}{\overset{\|}{C}}OH$

E $-NH-CH_2-\overset{OH}{\overset{|}{C}}-CH_2-NH-\overset{O}{\overset{\|}{C}}-CH_2-NH-CH_2-\overset{OH}{\overset{|}{C}}H-CH_2-NH_2$

F $-O-CH_2-\overset{OH}{\overset{|}{C}}H-CH_2-O-(CH_2)_4-O-CH_2-\overset{OH}{\overset{|}{C}}H-CH_2-$

G $-NH-CH_2-\overset{O}{\overset{\|}{C}}-NH-CH_2-\overset{O}{\overset{\|}{C}}-NH-CH_2-\overset{O}{\overset{\|}{C}}-OH$

Fig. 3.2 Commonly used spacer arms.

Several examples of commonly used spacer arms are shown in Fig. 3.2. Some of the earliest and most commonly used spacer arms are aminocarboxylic acids (A) (Cuatrecasas, 1970) and diamino alkanes (B, C) (Cuatrecasas, 1970). Spacer arms with terminal amino groups can be extended to give a longer spacer with a terminal carboxylic acid simply by reacting with succinic anhydride (Cuatrecasas, 1970). Spacer (D), which can be prepared in this manner, has a length of approximately 21 Å. Because of the problem of non-specific hydrophobic interactions, more hydrophilic spacer arms have recently been utilized. For example, spacer (E) can be prepared by reacting activated agarose alternately with diaminopropanol and the *N*-hydroxysuccinimide ester of bromoacetic acid (O'Cara, 1974). Long, uncharged hydrophilic spacers (F) result from activation of agarose using bisoxyranes (Sundberg and Porath, 1974). Oligopeptides such as glycyl-glycyl-glycine (G), glycyl-tyrosine, and glycyl-glycyl-cysteine also provide convenient commercially available hydrophilic spacers.

For certain applications it is advantageous to use macromolecular spacer arms such as polylysine, poly-(lysine-D L-alanine), or even naturally occurring proteins such as albumin as spacers (Parikh, *et al.,* 1974; Wilchek and Miron, 1974). These macromolecular spacers not only allow considerably greater separation of the attached ligand from the solid matrix with a minimal degree of hydrophobicity, but they allow multipoint attachment to the matrix, greatly reducing the likelihood of spontaneous leakage. If a naturally occurring protein such as albumin is used as a spacer it is advisable to couple it to the matrix in the presence of urea to promote coupling in the unfolded state, thus increasing the likelihood of multipoint attachment of the protein to the solid support.

3.2.4 Activation and coupling

Several excellent reviews have appeared recently that describe methods of activating matrices and covalently coupling ligands and spacers (Cuatrecasas, 1970; Parikh *et al.,* 1974; Porath and Axen, 1976). Furthermore, commercially available matrices that have preformed spacers and are activated and ready for use are becoming increasingly common. For these reasons this section will be limited. Only methods that can be applied with agarose matrices will be discussed. For methods that are particular to polyacrylamide, glass or cellulose the reader can refer to Inman (1974), Weetall and Filbert (1974) and Lilly (1976), respectively. Even for those methods that are discussed, no attempt will be made to describe the detailed chemical procedures involved. Instead, the general strategy of activation and coupling will be considered with emphasis on the uses and the advantages and disadvantages of some of the more common methods.

Two basic approaches are possible for coupling reactions. Derivatives of the ligand, containing spacers or activated groups, may be synthesized. These can then be directly coupled to the matrix. Alternatively, the matrix may be activated first and the desired spacers attached, and then the ligand coupled to it. The former approach permits more selective modification of the ligand and because the derivative can be purified and characterized, this approach gives a more uniform and well defined adsorbent. However, the chemistry involved is often quite difficult and will vary, depending upon the specific ligand. The second approach offers the convenience of solid phase synthesis. In addition, a number of standard procedures have been devised for activating agarose and synthesizing derivatives with various functional groups, obviating the need to develop special procedures for each new ligand. Fig. 3.3 illustrates several schemes for coupling ligands directly or via spacers.

Cyanogen bromide is the reagent most commonly used to activate agarose (Axen *et al.,* 1967). At high pH, cyanogen bromide reacts with hydroxyl groups in agarose to produce imidocarbonates (Porath and Axen, 1976) or cyanate esters (Kohn and Wilchek, 1978) that readily react with primary aliphatic or aromatic amines to give isourea derivatives. During activation protons are generated that must be neutralized either by titration with alkali or by performing the reaction in strong buffers. The buffer method is more convenient but the titration method can achieve higher degrees of substitution, and the progress of the reaction can be followed by the amount of alkali that must be added. The optimal conditions for activation and coupling using both methods have been discussed in detail (Cuatrecasas, 1970; March *et al.,* 1974). A simple colorimetric method for determining the degree of activation has been described (Kohn and Wilchek, 1978).

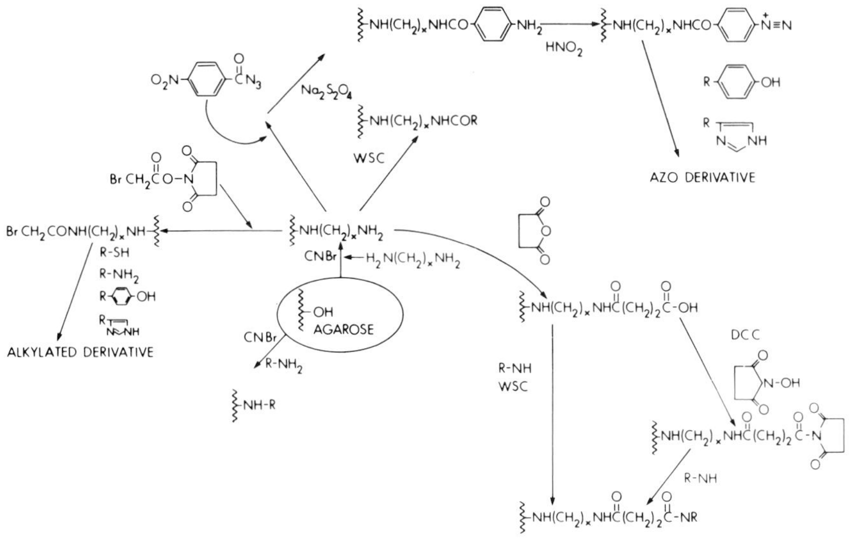

Fig. 3.3 Derivatives of agarose useful for the coupling of ligands.

The main disadvantage of cyanogen bromide is that the resulting linkage is somewhat unstable. The half-life of alanine-agarose prepared with cyanogen bromide is only 40 days in 0.05 M sodium bicarbonate, pH 8.0, at room temperature (Cuatrecasas and Parikh, 1972). This instability is partially due to the susceptibility of the isourea linkage to nucleophilic attack (Wilchek *et al.*, 1975; Tesser *et al.*, 1974). Therefore, buffers containing high concentrations of strong nucleophiles should be avoided for storage or operation of affinity adsorbents prepared with cyanogen bromide. If proper precautions are taken leakage will not seriously affect the usefulness of adsorbents prepared with cyanogen bromide, as evidenced by the wide success this method has had.

Reductive alkylation can be used to attach primary amines to agarose (Sanderson and Wilson, 1971). Vicinal hydroxyl groups are oxidized by sodium metaperiodate to aldehydes which then readily form Shiff bases with primary amines of the ligand. The linkage is stabilized by reduction of the Shiff bases with cyanoborohydride (Parikh *et al.*, 1974). Although the resulting linkage is quite stable, only terminal residues of agarose have vicinal hydroxyl groups. Therefore, the degree of coupling that can be achieved by activating agarose with sodium metaperiodate is much less than with cyanogen bromide. This is not true for polysaccharide matrices other than agarose.

Porath and co-workers have used several bifunctional reagents for coupling ligands to Sepharose (Porath and Axen, 1976). These reagents are of the general form XRY, where X and Y are reactive groups highly susceptible to nucleophilic attack. The reactive groups may be epoxides as in bisoxiranes (Sundberg and

Porath, 1974), activated double bonds as in divinylsulfone (Sairam and Porath, 1976) and benzoquinone (Brandt *et al.,* 1975), or activated halogens as in halohydrins (Sundberg and Porath, 1974), cyanuric chloride (Kay and Crook, 1967) and trifluoro-chloro-pyrimidine (Gribnau, 1977). Bisoxyranes (Sundberg and Porath, 1974) are an especially promising class of reagents. The resulting ether linkage should be very stable and is charge-free. In addition, a hydrophilic, non-ionic spacer is introduced during the process of activation. However, in general, experience with these bifunctional reagents for activating agarose has been limited compared with cyanogen bromide and in many cases the optimal conditions for coupling have not been determined. At the present time it is uncertain as to whether they will prove superior to cyanogen bromide.

The functional groups in those proteins and polypeptides that are available for coupling reactions are N-terminal α-amino groups, ε-amino groups of lysine, C-terminal carbonyl groups, carbonyl groups of aspartic and glutamic acid, sulfhydryl groups of cysteine, phenolic groups of tyrosine, imidazole groups of histidine, indole groups of tryptophane and aliphatic hydroxyl groups of serine and threonine. The functional groups most frequently used for coupling are amino groups. They readily react with agarose that has been activated using any of the methods described above. In addition they can be coupled to carbonyl derivatives of agarose via an amide bond. This can be accomplished using water soluble carbodiimides (Cuatrecasas, 1970). Alternatively, the stable activated *N*-hydroxysuccinimide ester derivative of the carbonyl group can be synthesized first. This readily reacts with amino groups in a second reaction (Cuatrecasas and Parikh, 1972). If the ligand to be coupled contains carbonyl groups in addition to amino groups, the *N*-hydroxysuccinimide method has the advantage that the complicating intra-ligand and inter-ligand reactions are avoided. Of course, carbodiimides also can be used to couple carbonyl-containing ligands to amino derivatives of agarose.

The sulfhydryl function is a more reactive nucleophile than the amino function and will take part in many of the reactions described above for amino groups. However, thioesters are considerably less stable than amides. Therefore, alkylating derivatives of agarose, such as bromoacetyl agarose (Cuatrecasas, 1970) or expoxy agarose (Sundberg and Porath, 1974), are preferable for coupling ligands through sulfhydryl groups. Bromoacetyl agarose will also react with amino, imidazole and phenolic functional groups.

The diazonium method (Cohen, 1974) provides a gentle and highly specific method for coupling proteins via phenolic (tyrosyl) or imidazole (histidyl) residues. The diazonium ion may also react with amino groups, guanidino groups (arginine) or indole groups (tryptophane) although these reactions are considerably slower than with phenol or imidazole. One feature of the azo-linkage that can be used to advantage is that it can be reduced readily with sodium dithionite, quantitatively releasing the coupled ligand. This permits elution of the intact receptor–ligand complex under relatively mild conditions.

Which specific group of the ligand is modified during coupling is critically important since attachment through certain groups can drastically reduce the ability of the ligand to bind to the receptor. Structure—activity relationships can be useful in predicting which modes of attachment will produce successful affinity adsorbents. For example, it is well known that tyrosyl residues of insulin are important for receptor binding and biological activity, but substitution of even rather bulky groups on the α-amino group of B1 or the ε-amino group of B29 has only minor effects on binding or activity (Blundell *et al.*, 1972). Consistent with this, insulin-agarose derivatives with tyrosine involved in an azo linkage are totally ineffective in purifying insulin receptors, whereas insulin-agarose derivatives coupled through the amino groups work well (Cuatrecasas, 1973).

One method to avoid the coupling of the ligand through an essential residue is illustrated by the coupling of lectins. Lectins are routinely coupled in the presence of the simple sugar to which they bind so that the sugar will mask the active site and stabilize the active conformation of the lectin during the coupling reaction (Cuatrecasas and Tell, 1973).

(a) *Blocking activated groups*
Frequently, activated groups are not completely consumed during the coupling reaction. Care should be taken to deactivate these groups. Their presence during subsequent fractionation procedures can cause covalent attachment of the substance to be purified or, more likely of more numerous impurities, resulting in the introduction of non-specific adsorbtion centers into the gel. Active groups can be blocked by incubating the gel with 0.1 M glycine or monoethanolamine. Monoethanolamine may be preferable because less ionic charge will be introduced.

(b) *The degree of substitution*
The concentration of bound ligand can have important and complicated effects on the binding strength and capacity of an affinity adsorbent (Harvey *et al.*, 1974). If the receptor is multivalent, high concentrations of coupled ligand can result in multipoint attachment of the receptor to the adsorbent and consequently very high avidity, whereas, with lower concentrations of coupled ligand, multipoint attachment is unlikely to occur. In addition, high concentrations of coupled ligand favor a retention effect (Silhavy *et al.*, 1975) due to the fact that it may be difficult for adsorbed receptor to escape from the immediate environment of the matrix, where the local ligand concentrations will be very high and diffusion may be restricted. This high avidity of binding is not always advantageous. It may make the subsequent elution of adsorbed receptor impossible except by conditions that result in its irreversible denaturation. Furthermore, for these same reasons, a high degree of substitution can greatly increase the avidity of non-specific binding. This is very clearly illustrated by the use of quaternary ammonium ions to purify acetylcholine receptors (Schmidt and Raftery, 1972). When a quaternary ammonium ligand was coupled to agarose to give a concentration of 2 mM, the column behaved as a weak ion exchanger with too much non-specific binding to be useful for purification of the receptor. However,

when the same ligand was coupled to give a concentration of 0.4 mM, its specificity was considerably increased.

As expected, increasing the coupled ligand concentration also tends to increase the capacity of the adsorbent. However, at high ligand concentrations, a further increase in coupled ligand will not result in a proportional increase in capacity, since there will be a growing number of inaccessible and therefore ineffective ligands with an increasing degree of substitution.

From the above considerations, it should be clear that decreasing the ligand concentration by diluting the affinity adsorbent with unsubstituted matrix, a procedure that will not affect the local concentration of coupled ligand, may not produce the same effects as preparing a homogeneously-substituted matrix with a lower coupled-ligand concentration.

The degree of substitution should be determined routinely. Usually, this is quite simple. The concentration of ligand in solution can be measured spectrophotometrically before and after coupling and the amount of ligand coupled determined by subtraction. When only a small fraction of the ligand present is coupled this method will be very inaccurate. In this case, a trace quantity of radioactive ligand can be included in the coupling reaction and the amount of radioactivity bound determined. Various other analytic methods can be used if neither of these methods are suitable (Gribnau, 1977; Lowe and Dean, 1974).

3.3 OPERATING CONDITIONS

3.3.1 Adsorption of receptor

(a) *Choice of buffer*

During the adsorptive phase of affinity chromatography, conditions should be chosen that maximize the amount of receptor adsorbed while minimizing the amount of impurities adsorbed. The range of pH, temperature, ionic strength and composition and detergent concentration that is permissible for ligand-receptor binding can usually be determined in preliminary studies using a soluble, labeled ligand. Non-specific ionic interactions can be minimized by increasing the ionic strength of the buffer. Hydrophobic interactions can be minimized at low ionic strength, by using chaotropic agents or organic solvents such as glycerol or ethylene glycol, and by using high detergent concentrations. In addition to preventing non-specific interactions with the affinity adsorbent, the presence of detergent is usually necessary to prevent aggregation of the receptor with other hydrophobic membrane proteins. For any particular application it is impossible to predict the optimal buffer composition. This should be varied systematically within the range permissible for binding until the most satisfactory results are obtained.

After the receptor is applied to the column, the column should be washed extensively to elute non-specifically adsorbed proteins. Often, once the receptor-ligand complex is formed, it is stable under buffer conditions that would prevent receptor binding from occurring. These harsh buffer conditions can be used to eliminate proteins that are non-specifically adsorbed. Since different kinds of non-specific binding can be inhibited by different kinds of buffers, it may be advantageous to wash the column with several different buffers. For example, protein non-specifically adsorbed by ionic interactions can be eluted with high ionic strength buffers. Later, proteins adsorbed by hydrophobic interactions can be eluted with buffers of low dielectric constants.

(b) *Batch verses column procedures*

Compared with batchwise techniques, column procedures for affinity chromatography offer certain theoretical advantages. Dilute samples can be concentrated in a narrow band near the top of the column and subsequently eluted in a sharp peak. Because of the concentrating effect of a column, the amount of protein adsorbed is usually independent of its concentration, while in a batchwise procedure the amount adsorbed from dilute solutions will be diminished (Lowe *et al.*, 1974). If several substances are adsorbed, resolution is often more satisfactory with a column, particularly if gradient elution is possible. Nevertheless, when using a very specific, very high affinity ligand, the two methods give comparable results. Batchwise procedures may be more convenient, particularly if small amounts of protein are to be adsorbed from large volumes of a crude sample that may cause a deterioration in flow rates in column procedures. Frequently, the adsorption phase is performed in a batchwise manner, the adsorbent is washed in a scintered glass funnel with vacuum, and then the slurry is poured into a column for elution.

(c) *Column geometry*

The effect of systematically varying column geometry on the results of affinity chromatography has been studied (Lowe *et al.*, 1974). Any change in geometry that increases the total bed volume increases the capacity of the column. If the column height is increased at the expense of the column diameter, so that the bed volume remains constant, the avidity of the column increases. These effects are most prominent at low concentrations of coupled ligand, and when the coupled ligand is of low affinity.

(d) *Flow rate*

For the maximal capacity of the column to be realized the flow rate must be sufficiently slow so that adsorption equilibrium between the immobilized ligand and the receptor is reached. The rate of association of the receptor with the coupled ligand will be considerably slower than with free ligand in solution. Diffusion of the coupled ligand is negligible by virtue of its attachment to the insoluble matrix. Furthermore, access of the receptor to the coupled ligand may be restricted by static films on the surface of the matrix and pores within the matrix. Protein impurities may diffuse into the pores and, for steric reasons, further restrict access of the receptor to the pores, thus further decreasing the rate of association.

The presence of receptor in several breakthrough fractions in a concentration less than that at which the receptor is being applied to the column suggests failure to attain adsorption equilibrium. If this occurs the breakthrough fractions can be recirculated through the column or the rate at which sample is applied to the column can be decreased. Alternatively, applying more dilute samples to the column, even at high flow rates, may prevent receptor from appearing in the breakthrough until the column approaches saturation (Cuatrecasas *et al.,* 1968).

3.3.2 Elution

Elution of adsorbed receptor may be accomplished by non-specific methods such as a change in pH, ionic strength, or temperature, or by including chaotropic agents, detergents or denaturing agents such as urea or guanidine in the elution buffer. Alternatively, a method that depends upon the biological specificity of the ligand—receptor interaction may be used. This is most commonly accomplished by including, in the elution buffer, a soluble counter ligand that will compete with the immobilized ligand. If immobilized lectins are used to purify glycoprotein receptors, simple sugars can be used as eluting agents. Here the eluting agent competes with the adsorbed receptor rather than with the immobilized ligand. This has the important advantage that the eluting agent (i.e. the sugar) will not interfere with the subsequent assay of the eluted receptor in the binding assay. Elution can also be accomplished by an allosteric inhibitor, or by omission or removal of an essential cofactor or activator of binding. For example, calcium-free buffer and EDTA have both been used to elute asialoglycoprotein receptors from asialo-orosomucoid agarose (Hudgin *et al.,* 1974), mannan-binding protein receptor from mannan agarose (Kawasaki *et al.,* 1978), intrinsic factor receptor from vitamin B12-intrinsic factor agarose (Marcoullis and Grasbeck, 1977) and transcobalamin II receptor from transcobalamin-vitamin B12-agarose (Seligman and Allen, 1978).

Biospecific elution is the more rational approach, and when successful it generally gives more satisfactory results. The reasons for this are clear. Biospecific elution depends upon the highly specific nature of the receptor—ligand interaction, whereas deforming buffers are as likely to elute contaminating proteins as they are the receptor. Furthermore, biospecific elution is much less likely to result in denaturation of the receptor. However, in practice biospecific elution of adsorbed receptor is frequently not possible. Several possible causes for this have already been discussed. If heavily substituted gels are used, multivalent binding or the retention effect may greatly increase the effective affinity of already high-affinity immobilized ligand—receptor interactions. In addition, once the receptor is bound to the immobilized ligand it may progressively become involved in non-specific interactions with the adsorbent. If this phenomenon, which has been called ligand-dependent non-biospecific adsorption (O'Cara, 1974), occurs, elution may be possible both by using a competitive ligand and by simultaneously altering the elution buffer to counter the non-specific effects. An example of this is provided by the purification of the β-adrenergic receptor on an alprenalol-agarose column (Vanquelin *et al.*, 1977). Although soluble alprenalol prevented the receptor from binding to the column, once bound it could be eluted by neither alprenalol nor sodium chloride alone. However, aprenalol in combination with 1 M sodium chloride eluted the receptor.

It is important to realize that, if biospecific elution is accomplished by a mobile competitive inhibitor, then unless there is a negatively cooperative interaction, the inhibitor will not actually displace the receptor from the immobilized ligand. Instead, the adsorbed receptor must first dissociate spontaneously from the immobilized ligand. The mobile inhibitor can then compete for rebinding to the immobilized ligand and elution will result. The spontaneous rate of dissociation of adsorbed receptor from high-affinity immobilized ligands such as polypeptide hormones may be very slow. For these systems, it may be necessary to clamp the column for up to several hours and possibly raise the temperature after applying the elution buffer to avoid excessively broad elution peaks.

3.4 PROBLEMS IN AFFINITY CHROMATOGRAPHY

3.4.1 Leakage

While affinity adsorbents are in general quite stable, a small fraction of immobilized ligand is inevitably released into the mobile phase. If immobilized ligands of relatively low affinity are used, the receptor will equilibrate rapidly between the mobile phase and the column. Therefore, even a few percent leakage will be of small consequence. However, if very high-affinity ligands are used, a small fraction of a percent leakage can seriously interfere with the performance of the affinity adsorbent. This is because the released ligand in the mobile phase is more accessible to the receptor, and it will have a higher probability of an initial interaction with the receptor while it is being applied to the column. Because the rate of dissociation of very high-affinity receptor–ligand complexes is slow on the time scale of the affinity chromatography procedure, for practical purposes the receptor will be tied up irreversibly by the released ligand. In such cases the presence in breakthrough effluents of ligand that has leaked from the column, in addition to preventing adsorption of the receptor to the column, may interfere with the subsequent assay of receptor that has passed through the column without being adsorbed. This may lead to the erroneous conclusion that irreversible adsorbtion to the gel has occurred.

Several measures can be taken to minimize the problem of leakage. In high-affinity systems, the affinity adsorbent used should contain the least quantity of ligand substitution compatible with effective adsorption. High concentrations of coupled ligand generally result in the release of proportionately greater quantities of free ligand. Excess bound ligand will not improve the performance of the column, but the resulting increase in free ligand may paradoxically decrease the amount of receptor adsorbed. This is particularly true if a high degree of ligand substitution results in bound ligand that is not accessible for receptor binding because of steric constraints, as discussed in previous sections.

Frequently the chemical methods used to immobilize the ligand result in several different modes of attachment, and only a fraction of these may retain the activity of the free ligand. This situation may not appear detrimental, as long as sufficient ligand can be coupled in the active form so that the adsorbent has sufficient affinity and capacity. However, when inactive species are released into the mobile phase they may regain activity, adding to the problem of leakage without enhancing the capacity or affinity of the adsorbent. Therefore, to minimize leakage, a coupling method should be used that results in a well characterized, homogeneous mode of attachment that preserves the highest degree of ligand activity possible.

The lability of the cyanogen bromide linkage is a source of leakage that has been discussed in a previous section. Since the isourea linkage that results from the coupling of amino groups to cyanogen bromide-activated agarose is subject to nucleophilic attack (Wilchek *et al.*, 1975; Tesser *et al.*, 1974), buffers containing high concentrations of strong nucleophiles should be avoided. Newer methods of activating agarose may result in more stable linkages. This has not been extensively studied, however. The products of divinyl sulfone activation are very labile in alkaline solutions, making it likely that leakage will be a problem (Sairam and Porath, 1976).

Agarose itself has a low degree of solubility and this may contribute somewhat to leakage, particularly if extremes of pH, elevated temperatures, or denaturants such as urea are used (Porath and Kristiansen, 1975). This component of the problem is greatly reduced by the chemically crosslinking agarose (see Section 3.2.1).

Leakage can also be diminished by using soluble macromolecular spacers to separate the gel matrix from the immobilized ligand (Parikh *et al.*, 1974; Wilchek and Miron, 1974). Since macromolecules such as the branched copolymer, polylysine-polyalanine, can attach to agarose at multiple points, the overall chemical stability of the linkage will be greatly enhanced.

Frequently the greatest source of leakage is the release of material that is non-specifically adsorbed to the gel, rather than the release of convalently bound material by cleavage of chemical bonds. Several substances such as insulin and glucagon adsorb tenaciously to agarose and other matrices and their removal may require days of washing with high ionic strength buffers and urea (Parikh *et al.*, 1974). The need for extensive washing to remove adsorbed ligand after the preparation of the gel is generally appreciated. However, the continual leakage of covalently bound ligand during storage can lead to renewed build-up of adsorbed material. Therefore, after storage, the affinity adsorbent should again be washed extensively before it is used.

3.4.2 Non-specific interactions

The problem of non-specific adsorption (i.e. adsorption due to gross physical—chemical interactions not based on biological specificity) has been extensively reviewed (O'Cara *et al.,* 1974). These effects may not only decrease the specificity of an affinity adsorbent, thus limiting the degree of purification that is possible, but they may also interfere with the elution of adsorbed receptor.

Agarose, particularly after removal of its sulfate groups, is virually free of non-specific adsorption centers. Non-specific adsorption centers are introduced by the isourea function which results from the coupling of aliphatic amines with cyanogen bromide-activated agarose. These isourea groups are fully protonated at neutral pH and will interact with negatively charged groups (Jost *et al.,* 1974; Wilchek and Miron, 1974). This ion-enchange effect can be eliminated by coupling ligands via a hydrazide group. The resulting agarose derivatives have a pK_a of 4.2 (Jost *et al.,* 1974) and therefore are uncharged in the pH range at which they are usually used. The most serious non-specific adsorption effects usually result from long hydrophobic spacers (O'Cara *et al.,* 1974). Nonpolar, hydrophilic spacers should be used where possible (see Section 3.2.3). The ligand, itself, can be the source of non-specific interactions as illustrated by the use of quaternary ammonium compounds to purify the acetylcholine receptor (Schmidt and Raftery, 1972) or when 'sticky' polypeptides such as insulin are immobilized.

The forces involved in non-specific interactions are individually quite weak, and become significant only if multiple points of binding occur. A sufficient distance between substituted groups on the surface of the matrix makes it unlikely that a particular protein will interact simultaneously with more than one immobilized group. Therefore the degree of substitution of the matrix will greatly affect its non-specific adsorption properties.

When derivatives of agarose (e.g., diamino-dipropylamino-agarose or succinyl-agarose) are used to couple ligands, the coupling reaction will often not quantitatively consume all the derivative groups. These unreacted groups contribute to the non-specific properties of the affinity adsorbent. Their number should be kept to a minimum by preparing primary derivatives of agarose with the least degree of substitution necessary for the subsequent coupling of sufficient ligand. Similarly, if the chemical method of coupling the ligand results in a mixed mode of attachment of the ligand, only a small fraction of the attached ligand may retain its high affinity. However, that fraction of the immobilized ligand that is inactive is just as likely to contribute to the non-specific properties of the adsorbent as the fully active immobilized ligand. Therefore, highly specific chemical methods of immobilizing the ligand are preferable.

3.5 INDIRECT AFFINITY CHROMATOGRAPHY

A major obstacle to the general applicability of affinity chromatography to the purification of membrane receptors is that, after being solubilized, many membrane receptors lose their binding ability. Nevertheless, with certain modifications of the usual technique, affinity chromatography may be useful for the purification of even these receptors.

Usually, in affinity chromatography, the immobilized ligand is coupled to the matrix covalently. Non-covalent attachment is also possible, providing the link is of sufficient strength. For example, it may be accomplished by immobilizing antibodies to the ligand. The ligand can be coupled by virtue of its tight binding to the antibodies. It may also be accomplished by the convalent coupling of biotin to the ligand. The biotinyl ligand can then be coupled non-covalently to immobilized avidin. Because of the very high affinity of the biotin–avidin interaction ($k_{diss} \sim 10^{-15}$ M), this method will result in a non-convalent attachment that is virtually as stable as a covalent bond. Biotin derivatives of ACTH (Hofmann and Kiso, 1976) and insulin (Hofmann *et al.*, 1977; May *et al.*, 1978) have been prepared which, after coupling to avidin-agarose, retain their affinities for their respective receptors. Biotin-insulin immobilized on avidin-Sepharose has been used to adsorb insulin receptors (Bayer and Wilchek, 1978).

Non-covalent coupling of the ligand to the solid support makes it possible to incubate the free, soluble ligand with the receptor first, and then to adsorb the receptor–ligand complex to the solid support (Fig. 3.4). The analogy to indirect immunopercipitation or indirect immunocytochemistry is obvious. As an example of this technique, we have incubated biotin-insulin with soluble insulin receptors, adsorbed the complex to avidin-Sepharose, and eluted insulin receptors with urea. This indirect method has no obvious advantage over conventional affinity adsorbents with insulin covalently coupled to agarose, but for certain receptors, such as that for glucagon, it may be the only method that works. After glucagon receptors are solubilized from the membrane they are no longer able to bind glucagon (Giorgio *et al.*, 1974). However, if glucagon is bound to the receptor prior to solubilization, the complex is stable after solubilization, and therefore may possibly be purified by this indirect technique.

A logical extension of this technique is to covalently couple the ligand to the membrane-bound receptor using an affinity label or bifunctional crosslinking reagent, to solubilize the receptor–ligand complex, and to then non-convalently adsorb the ligand, with receptor bound to it, to the solid support. While this kind of technique has been used extensively for the purification of affinity-labeled peptides (Wilchek, 1974), there are only two examples of its use for the purification of membrane receptors. One is the purification of a biotin binding and transport protein from yeast using a biotin affinity label and avidin-Sepharose (Bayer *et al.*, 1976), and the purification of epidermal growth factor receptors using a photo-affinity labeling derivative of EGF and anti-EGF antibody-Sepharose (Hock *et al.*, 197

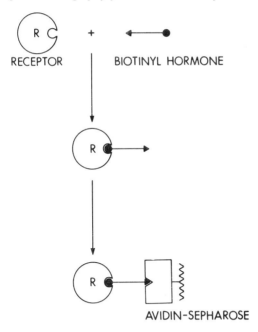

Fig. 3.4 Indirect affinity chromatography utilizing the biotin-avidin system.

This type of method, because it does not require that the receptor retain its binding activity after solubilization, has the potential of greatly increasing the number of receptors that will eventually be purified by affinity chromatography. (Recently this approach has also been used to purify insulin receptors (Heinrich *et al.*, 1980)).

REFERENCES

Aharonov, A., Abramsky, O., Tarrab-Hazdai, R. and Fuchs, S. (1975), *Lancet*, **1**, 340–342.

Aharonov, A., Tarrab-Hazdai, R. and Silman, J. and Fuchs, S. (1977), *Immuno-chemistry*, **14**, 129–137.

Almon, R., Andrew, C. and Appel, S. (1974), *Science*, **186**, 55–57.

Axen, R., Porath, J. and Ernback, S. (1967), *Nature*, **214**, 1302–1304.

Bayer, E.A., Viswamatha, T. and Wilcheck, M. (1976), *FEBS Letters*, **60**, 309–312.

Bayer, E.A. and Wilchek, M. (1978), *TIBS*, **3**, N257–N259.

Blundell, T.L., Hodgkin, D.C., Dodson, G.C. and Mercola, D.A. (1972), *Adv. prot. Chem.*, **26**, 280–294.

Brandt, J., Anderson, L.-O. and Porath, J. (1975), *Biochem. biophys. Acta*, **386**, 196–202.

Brockes, J.P. and Hall, Z.W. (1975), *Proc. natn. Acad. Sci. U.S.A.*, **72**, 1368–1372.

Caron, M.G., Srinevasa, Y., Pitha, J., Kocilek, K. and Lefkowitz, R.J. (1979), *J. biol. Chem.*, **254**, 2923–2927.

Chang, H.W. (1974), *Proc. natn. Acad. Sci., U.S.A.,* **71**, 2113–2117.

Cohen, L.A. (1974), In: *Methods in Enzymology,* (Jakoby, W.B. and Wilchek, M., eds.), Vol. XXXIV, Part B, Academic Press, New York, pp. 102–108.

Conrad, D.H. and Froese, A. (1978), *J. Immunol.,* **120**, 429–437.

Cooper, H.A., Clemetson, K.J. and Luscher, E.F. (1979), *Proc. natn. Acad. Sci., U.S.A.,* **76**, 1069.

Cresswell, P. (1979), *J. biol. Chem.,* **254**, 414–419.

Cuatrecasas, P. (1970), *J. biol. Chem.,* **245**, 3059–3065.

Cuatrecasas, P. (1973), *J. biol. Chem.,* **248**, 3528–3534.

Cuatrecasas, P. and Anfinsen, C.B. (1971), *Ann. Rev. Biochem.,* **40**, 259–279.

Cuatrecasas, P. and Parikh, I. (1972), *Biochemistry,* **11**, 2291–2299.

Cuatrecasas, P. and Tell, G.P.E. (1973), *Proc. natn. Acad. Sci., U.S.A.,* **70**, 485–489.

Cuatrecasas, P., Tell, G.P. E., Sica, V., Parikh, I. and Chang, K.-J. (1974), *Nature,* **247**, 92–97.

Cuatrecasas, P., Wilchek, M. and Anfinsen, C. (1968), *Proc. natn. Acad. Sci., U.S.A.,* **61**, 636–643.

Dufau, M.L., Ryan, D., Baukal, A. and Katt, K.J. (1975), *J. biol. Chem.,* **250**, 4822–4824.

Er-el, Z., Zaidenzaig, Y. and Shaltiel, S. (1972), *Biochem. biophys. res. Commun.,* **49**, 383–390.

Flier, J.S., Kahn, C.R., Roth, J. and Bar, R.S. (1975), *Science,* **190**, 63–65.

Giorgio, N.A., Johnson, C.B. and Blecher, M. (1974), *J. biol. Chem.,* **249**, 428–437.

Gribnau, T.C.J. (1977), *Coupling of Effector-Molecules to Solid Supports,* Drupperij van Mameren, B.V., Nijmegen, The Netherlands.

Harvey, M.J., Lowe, C.R., Craven, D.B. and Dean, P.D.G. (1974), *Eur. J. Biochem.* **41**, 335–340.

Heinrich, J., Pilch, P.F. and Czech, M.P. (1980), *J. biol. Chem.,* **255**, 1732–1737.

Hock, R.A., Nexo, E. and Hollenberg, M.D. (1979), *Nature,* **277**, 403–405.

Hofmann, K., Finn, M.F., Friesen, H.J., Diaconescu, C. and Zahn, H. (1977), *Proc. natn. Acad. Sci., U.S.A.,* **74**, 2697–2700.

Hofmann, K. and Kiso, Y. (1976), *Proc. natn. Acad. Sci., U.S.A.,* **73**, 3516–3518.

Hofstee, B.H.J. (1973), *Analyt. Biochem.,* **52**, 430–448.

Homcy, C.J., Wrenn, S.M. and Haber, E. (1977), *J. Biol. Chem.,* **252**, 8957–8964.

Hudgin, R.L., Pricer, W.E., Jr., Ashwell, G., Stockert, R.J. and Morell, A.G. (1974), *J. Biol. Chem.,* **249**, 5536–5543.

Inman, J.K. (1974), In: *Methods in Enzymology,* (Jakoby, W.B. and Wilchek, M., eds.), Vol. XXXIV, Part B, Academic Press, New York, pp. 30–58.

Inman, J.K. and Dintzis, H.M. (1969), *Biochemistry,* **8**, 4074–4082.

Jacobs, S., Chang, K.-J. and Cuatrecasas, P. (1978), *Science,* **200**, 1283–1284.

Jacobs, S., Shechter, Y., Bissell, K. and Cuatrecasas, P. (1977), *Biochem. biophys. res. Commun.,* **77**, 981–988.

Jost, R., Miron, T. and Wilcheck, M. (1974), *Biochim. biophys. Acta,* **362**, 75–82.

Karlin, A. and Cowburn, D. (1973), *Proc. natn. Acad. Sci., U.S.A.,* **70**, 3636–3640.

Karlsson, E., Heilbronn, E. and Widlund, L. (1972), *FEBS Letters,* **28**, 107–111.

Kawasaki, T., Etoh, R. and Yamashina, I. (1978), *Biochem. biophys. res. Commun.*, **81**, 1018–1024.

Kay, G. and Crook, E. (1967), *Nature*, **216**, 514–515.

Kohn, J. and Wilchek, M. (1978), *Biochem. biophys. res. Commun.*, **89**, 7–14.

Kouvonen, I. and Grasbeck, R. (1979), *Biochem. biophys. res. Commun.*, **86**, 358–364.

Kristiansen, T. (1974), In: *Methods in Enzymology*, (Jakoby, W.B. and Wilchek, M., eds.), Vol. XXXIV, Part B, Academic Press, New York, pp. 331–341.

Lilly, M.D. (1976), In: *Methods in Enzymology* (Mosbach, K., ed.), Vol. XLIV, Academic Press, New York, pp. 46–53.

Lindstrom, J., Einarson, B. and Merlie, J. (1977), *Proc. natn. Acad. Sci., U.S.A.*, **75**, 769–773.

Lindstrom, J., Lennon, V., Seybold, M. and Wittingham, S. (1976), *N.Y. Acad. Sci.*, **274**, 254–274.

Lowe, C.R. and Dean, P.D.G. (1974), *Affinity Chromatography*, John Wiley and Sons, London.

Lowe, C.R., Harvey, M.J., Craven, D.B. and Dean, P.D.G. (1973), *Biochem. J.*, **133**, 499–506.

Lowe, C.R., Harvey, M.J. and Dean, P.D.G. (1974), *Eur. J. Biochem.*, **41**, 341–345.

McIntosh, C., Warnecke, J., Nieger, M., Barner, A. and Kobberling, J. (1976), *FEBS Letters*, **66**, 149–154.

May, J.M., Williams, R.H. and De Haen, C.D. (1978), *J. biol. Chem.*, **253**, 686–690.

March, S.C., Parikh, I. and Cuatrecasas, P. (1974), *Anal. Biochem.*, **60**, 149–152.

Marcoullis, G. and Grasbeck, R. (1977), *Biochem. biophys. Acta*, **499**, 309–314.

Merlies, J.P., Sobel, A., Changeux, J-P. and Gros, F. (1975), *Proc. natn. Acad. Sci., U.S.A.*, **72**, 4028–4032.

O'Cara, P. (1974) In: *Industrial Aspects of Biochemistry*, (Spenser, B., ed.), North-Holland Publishers, Amsterdam, pp. 107.

O'Cara, P., Barry, S. and Griffin, T. (1974), In: *Methods in Enzymology*, (Jakoby, W.B. and Wilchek, M., eds.), Academic Press, New York, pp. 108–126.

Olsen, R.W., Meunier, J-C. and Changeux, J-P. (1972), *FEBS Letters*, **28**, 96–100.

Onj, D.E. and Brady, R.N. (1974), *Biochemistry*, **13**, 2822–2827.

Porath, J. and Axen, R. (1976), In: *Methods in Enzymology*, (Mosback, K., ed.), Vol. XLIV, Academic Press, New York, pp. 19–45.

Porath, J., Janson, J.-C. and Laas, T. (1971), *J. Chromatogr.*, **60**, 167–177.

Porath, J. and Kristiansen, T. (1975), In: *The Protein*, (Neurath, H. and Hill, R.L., eds.), 3rd Ed., Vol. I, Academic Press, New York, pp. 95–178.

Porath, J., Laas, T. and Janson, J.-C. (1975), *J. Chromatogr.*, **103**, 49–62.

Parikh, I. and March, S. and Cuatrecasas, P. (1974), In: *Methods in Enzymology*, (Jakoby, W.B. and Wilchek, M., eds.), Vol. XXXIV, Part B, Academic Press, New York, pp. 77–102.

Patrick, J., Lindstrom, J., Culp, B. and McMiland, J. (1973), *Proc. natn. Acad. Sci., U.S.A.*, **70**, 3334–3338.

Rees, D.A. (1969), *Advan. Carbohyd. Chem.*, **25**, 321–

Sairam, M.R. and Porath, J. (1976), *Biochem. biophys. res. Commun.*, **69**, 190–196.

Sanderson, C.J. and Wilson, D.V. (1971), *Immunology*, **20**, 1061.

Schmidt, J. and Raftery, M.A. (1972), *Biochem. biophys. res. Commun.*, **49**, 572–578.

Seligman, P.A. and Allen, R.H. (1978), *J. biol. Chem.,* **253**, 1766–1772.

Shaltiel, S. (1974), In: *Methods in Enzymology,* (Jakoby, W.B. and Wilchek, M., eds.), Vol. XXXIV, Part B, Academic Press, New York, pp. 126–140.

Shiu, R.P.C. and Friesen, H.G. (1975), *J. biol. Chem.,* **249**, 7902–7911.

Shiu, R.P.C. and Friesen, H.G. (1976), *Science,* **192**, 259–261.

Silhavy, T.J., Szmeleman, S., Boos, W. and Schwartz, M. (1975), *Proc. natn. Acad. Sci., U.S.A.,* **72**, 2120–2124.

Smith, B.R. and Hall, R. (1974), *Lancet,* **2**, 427–430.

Steers, E. Jr., Cuatrecasas, P. and Pollard, H.B. (1971), *J. biol. Chem.,* **246**, 196–200.

Sundberg, L. and Porath, J. (1974), *J. Chromatogr.,* **90**, 87–98.

Tate, R.L., Holmes, J.A.M., Kohn, L.D. and Winand, R.J. (1975), *J. biol. Chem.,* **250**, 6527–6533.

Tesser, G.I., Fisch, H.-U. and Schwyzer, R. (1974), *Helv. Chim. Acta,* **57**, 1718–1730

Valderama, R., Weill, C., McNamee, M.G. and Karlin, A. (1976), *Proc. natn. Acad. Sci., U.S.A.,* **274**, 108–115.

Vanquelin, G., Geynet, P., Hanoune, J. and Strosberg, A.D. (1977), *Proc. natn. Acad. Sci., U.S.A.,* **74**, 3710–3714.

Waters, M.J. and Friesen, H.G. (1978), *Proc. Aust. Endocr. Soc.,* **21**, 45.

Weetall, H.H. and Filbert, A.M. (1974), In: *Methods in Enzymology,* (Jakoby, W.B. and Wilchek, M., eds.), Vol. XXXIV, Part B, Academic Press, New York, pp. 59–72.

Wilchek, M. (1974), In: *Methods in Enzymology,* (Jakoby, W.B. and Wilchek, M., eds.), Vol. XXXIV, Part B, Academic Press, New York, pp. 182–195.

Wilchek, M. and Miron, T. (1974), In: *Methods in Enzymology,* (Jakoby, W.B. and Wilchek, M., eds.), Vol. XXXIV, Part B, Academic Press, New York, pp. 72–76.

Wilchek, M., Oka, T. and Topper, Y.J. (1975), *Proc. natn. Acad. Sci., U.S.A.,* **72**, 1055–1058.

Yon, R.J. (1972), *Biochem. J.,* **126**, 765–767.

4 Affinity Labeling of Hormone Receptors and Other Ligand Binding Proteins

PETER S. LINSLEY, MANJUSRI DAS
and
C. FRED FOX

Acknowledgements
This work was supported by research grants from the American Cancer Society
(BC-79), from the United States Public Health Services (AM-25826), and by project
funds from a Muscular Dystrophy Association grant to the Jerry Lewis Neuromuscular
Research Center. The procedure for synthesis of 4, 4' dithiobisphenylazide-*S, S'*-
dioxide and PAPDIP were developed with the technical assistance of Pam Billings.
Peter Linsley is a predoctoral trainee supported by a National Research Service
Award in Tumor Cell Biology (CA-09056).

Membrane Receptors : *Methods for Purification and Characterization*
(*Receptors and Recognition,* Series B, Volume 11)
Edited by S. Jacobs and P. Cuatrecasas
Published in 1981 by Chapman and Hall, 11 New Fetter Lane, London EC4P 4EE
© Chapman and Hall

4.1 INTRODUCTION

This chapter describes the use of cross-linking technology to identify ligand-binding receptors of biological interest. We have updated a recent review from this laboratory on cross-linking (Das and Fox, 1979), and have added to it two additional components. First, we describe a simple synthetic procedure for the preparation of methyl[3-(p-azidophenyl)-dithio]-propionimidate, PAPDIP, a reversible, hetero-bifunctional cross-linking reagent (see Table 4.1). Second, we described a reaction which results in the spontaneous 'direct linkage' of some radioiodinated biological ligands to their receptors without the use of chemical cross-linking. This direct linkage reaction has now been described for the addition of EGF to its receptors on a variety of cultured cells and isolated membranes, and for the addition of thrombin to a receptor protein on certain cultured cell lines. The chapter concludes with a review of recent studies which have exploited covalent EGF–EGF receptor complexes to map the position of regions of biochemical interest on the receptor protein.

4.2 PHOTOREACTIVE CROSS-LINKING REAGENTS

The investigator seeking to determine the mechanism of cellular response to a particular ligand–receptor interaction faces a formidable challenge. Cells are complex multicomponent systems, and the classical biochemical approach of isolation and characterization of protein components does not always reveal the transient interactions that occur in the natural environment of the cell. One promising strategy for visualizing these interactions is stabilization by covalent chemical cross-linking.

The covalent cross-linking of a radioactive ligand to its binding protein(s) is a derivitive of chemical cross-linking technology known as 'affinity labeling'. The cross-linking reagents most useful for affinity labeling are the photoactive arylazides which have the convenient property of becoming reactive only after photolysis. A number of useful photoactive arylazides are described in Table 4.1. An increasing number of these reagents is becoming commercially available

In contrast to the more commonly used site-specific cross-linking reagents, the arylazides can react with components in a nonpolar, hydrophobic environment. Arylazides are photolyzed at wavelengths (300–400 nm) which do not cause direct photochemical damage to proteins or nucleic acids. Arylnitrenes generated by photolysis of arylazides have lifetimes on the order to 10^{-4}s (Reiser et al.,1968). The various reactions open to an arylnitrene have been discussed by Knowles (1972) and by Peters and Richards (1977). A nitrene can react with C–H bonds by

Table 4.1 Photoactivable, bifunctional cross-linking reagents

Cross-linker (Ref.)	Cleavability of cross-link	Formula	Site specificity of the non-photoactivable group
4-Fluoro-3-nitrophenylazide Fleet et al., 1972	—		Amino
1,5-Diazidonaphthalene Mikkelson and Wallach, 1976	—		—
4,4'-Diazidobiphenyl Mikkelson and Wallach, 1976	—		—
4,4'-Dithiobisphenylazide Mikkelson and Wallach, 1976	+		—
N,N'-(2-nitro-4-azidophenyl) cystamine-S,-S'-dioxide Huang and Richards, 1977	+		Sulfhydryl

(continued on the next page)

Table 4.1 Photoactivable, bifunctional cross-linking reagents (*continued*)

Cross-linker (Ref.)	Cleavability of cross-link	Formula	Site specificity of the non-photoactivable group
p-Azidophenacyl bromide Hixon and Hixon, 1975	—	N_3-⟨benzene⟩-$C(=O)-CH_2-Br$	Sulfhydryl
Methyl-4-azidobenzoimidate Ji, 1977	—	N_3-⟨benzene⟩-$C(=NH)-OCH_3$	Amino
Ethyl(4-azidophenyl)-1, 4-dithiobutyrimidate Kiehn and Ji, 1977	+	N_3-⟨benzene⟩-$S-S-(CH_2)_3-C(=NH)-O-CH_2-CH_3$	Amino
Methyl [3-(*p*-azidophenyl)- dithio] -propionimidate Das *et al.*, 1977; Das *et al.*, 1978; Das and Fox, 1978	+	N_3-⟨benzene⟩-$S-S-(CH_2)_2-C(=NH)-O-CH_2-CH_3$	Amino
4-Azidobenzoyl-*N*-hydroxysuc- cinimide Yip *et al.*, 1980	—	N_3-⟨benzene⟩-$C(=O)-O-N$⟨succinimide⟩	Amino

(*continued on the next page*)

Table 4.1 Photoactivable, bifunctional cross-linking reagents (*continued*)

Cross-linker (Ref.)	Cleavability of cross-link	Formula	Site specificity of the non-photoactivable group
N-succinimidyl-6-4′-azido-2′-nitrophenylamino)hexanoate Hock *et al.*, 1979	—		Amino
N-(5-azido-2-nitrobenzoyloxy) succinimide Lewis *et al.*, 1977	—		Amino
N-(5-azido-2-nitrophenyl) ethylenediamine Darfter and Marinetti, 1977	—		Carboxyl
N-(4-azido-2-nitrophenyl) ethylenediamine Lee *et al.*, 1979; Carney *et al.*, 1979; Das and Fox, 1978	—		Carboxyl

abstraction and insertion, and may therefore create a cross-link in the absence of any particular functional group on a protein.

The heterobifunctional reagents, which attach initially by normal chemical modification, and secondarily by photolysis of an appropriate group, are particularly useful for identifying ligand-binding components in complex biological systems. One such reagent is methyl[3-(*p*-azidophenyl)-dithio]-propionimidate (PAPDIP). Since this reagent is not commercially available, we describe its synthesis here.

In our initial studies on photoaffinity labeling of EGF receptors, we employed PAPDIP synthesized by a relatively complex multistep procedure, essentially identical to that described by Kiehm and Ji (1977) for the synthesis of ethyl(4-azidophenyl)-1,4-dithiobutyrimidate. This procedure was never published because it did not produce a product of sufficient purity to allow characterization. More recently, we have used different procedures for the syntheses of PAPDIP. In one procedure, we reacted the photoreactive *S-S '*-dioxide described here (compound II) with 3-thiopropionitrile and converted the mixed disulfide product to PAPDIP by the standard treatment with HCl in anhydrous methanol (Pinner, 1883). The product synthesized by this procedure appeared pure, but it underwent disulfide exchange to yield a mixture of PAPDIP and the two possible homobifunctional products, 4,4 '-dithiobisphenylazide (Mikkelson and Wallach, 1976) and dithiobispropionimidate (Wang and Richards, 1974). The formation of these products from PAPDIP preparations which initially appear to be pure is noticeable within two to three days of PAPDIP synthesis, and equilibrium is generally reached within a week to ten days. We are yet to synthesize a PAPDIP preparation which has remained a single component for more than a few days. Nevertheless, the other procedure for synthesis of PAPDIP described here is of sufficient simplicity to obviate the problem. It can be completed within minutes and uses precursors that have a long shelf life. This synthesis exploits the reaction of an alkylthiol with 4,4 '-dithiobisphenylazide-*S, S '*-dioxide, prepared essentially as described by Huang and Richards for the synthesis of *N-N '*-(2-nitro-4-azidophenyl)cystamine-*S, S '*-dioxide (Huang and Richards, 1977).

(a) *Synthesis of 4,4 '-dithiobisphenylazide-S-S '-dioxide*

4,4 '-dithiobisphenylazide (I)   4,4 '-dithiobisphenylazide -
                                                  S-S '-dioxide (II)

10 g (32 mmol) of (I) (Mikkelson and Wallach, 1976; also available commercially from Pierce Chemical Company) was dissolved in 1500 ml of chloroform at room temperature. Although (I) is soluble in far less chloroform, peracetic acid is not; the purpose of the large volume of chloroform is to accomodate the peracetic acid.

10 ml of 40% peracetic acid was added dropwise to the solution of (I) in chloroform. The reaction mixture was stirred for 4 h at $0°C$ and then for an additional hour at room temperature. These and all subsequent steps were performed in the dark using a photographic safelight for illumination. The reaction mixture was concentrated under vacuum by rotary evaporation at $25-30°C$, and the product was suspended at room temperature in the minimal volume of chloroform (approximately 130 ml). The chloroform solution was stored in the freezer overnight to promote crystallization. The reaction product was filtered on Whatman I paper, yielding 9.2 g of green, cuboidal crystals. These were slightly sticky, and the product was dissolved in the minimal volume of chloroform at room temperature and recrystallized twice. The final product (yield, 5.8 g) produced a single spot on silicic acid thin layer chromatography developed either with chloroform (R_F = 0.61) or 1:1 chloroform: methanol (R_F = 0.80). This product has been stored for over 18 months at $-20°C$ with no signs of deterioration.

(b) *Synthesis of PAPDIP*

In this synthetic procedure, 3-thiopropionimidate (Traut *et al.*,1973) was reacted directly with anhydrous (II) in anhydrous methanol. Reactant (II) was maintained in a vacuum over P_2O_5 to assure dryness. When equimolar quantities of the photo-reactive thiosulfonate (II) and 3-thiopropionimidate were mixed, (10 μmol of each in 1.0 ml) the reaction proceded quantitatively in 10 minutes at room temperature, yielding PAPDIP and the other product of the reaction, a sulfinic acid which does not react with proteins. The PAPDIP formed had an R_F of less than 0.1 after silicic acid thin layer chromatography in chloroform, permitting easy separation from the components of the reaction mixture and assessment of the progress of the reaction. The PAPDIP synthesized by this reaction was also unstable, and other disulfides, formed through exchange, were detected within a few days of storage. PAPDIP synthesized in this fashion should be employed as soon as possible after synthesis for optimal use as a photoaffinity reagent.

4.3 PHOTOREACTIVE DERIVITIVES OF LIGANDS

The reagents listed in Table 4.1 can be used to construct photoreactive derivatives of polypeptide ligands for affinity labeling studies. The properties of a number of these have been reviewed (Das and Fox, 1979) and are summarized in Table 4.2. The use of photoreactive compounds to construct affinity labeling reagents is illustrated by the following recent examples.

Yip *et al.*, have reported the synthesis and characterization of photoreactive insulins (1980). The photoreactive derivatives $N^{\epsilon B29}$-(azidobenzoyl) insulin (MAB-insulin) and $N^{\alpha A1}$, $N^{\epsilon B29}$-di(azidobenzoyl) insulin (DAB-insulin) were synthesized by reacting bovine insulin with an amino group specific reagent, 4-azidobenzoyl-N-hydroxysuccinimide. These derivatives were purified by ion-exchange chromatography, and their identities established by amino acid analysis and end-group determination. The photoreactivities of the two derivatives were then demonstrated by their spectral properties and by their ability to form covalent polymers of high molecular weight when exposed to light. The apparent binding affinities for DAB-insulin and MAB-insulin were 1.4×10^{-10} M and 3.8×10^{-11} M respectively, compared to 2.7×10^{-11} M for underivatized insulin. The loss of biological activity in DAB-insulin where the α-amino group of gly A−1 was modified is consistent with the important role of this residue in insulin action.

Two recent reports describe the construction of photoactive derivities of the enkephalins − a class of endogenous, opiate-like neurotransmitters. Hazum *et al.*, (1979) described the synthesis and binding properties of D-Ala^2, Leu^5 enkephalin-N-(2-nitro-4-azidophenyl)Lys. Another arylazido-enkephelin derivative, [D-Ala^2-Met^5]-enkephalin-Tyr-N-(2-nitro-4-azidophenyl)ethylenediamine (ETN) was synthesized by Lee *et al.* (1979). Initially, tyrosine was condensed with N-(2-nitro-4-azidophenyl)ethylenediamine (NAPEDE), yielding Tyr-NAPEDE. The amino group in Tyr-NAPEDE was then condensed with the C-terminal carboxyl group in [D-Ala^2-Met^5]-enkephelin, to yield the final product. The extra tyrosine residue in this derivative was added to provide a site for radioiodination, since in some experiments the iodination of the N-terminal tyrosine in the enkephalin appeared to lead to inactivation of receptor binding activity. The derivative synthesized by Lee *et al.* inhibited the binding of [^{3}H]-enkephalinamide to enkephalin receptor-rich NG-108 cell membranes with an I_{50} of 22 nM. Photolysis of membranes in the presence of ETN caused irreversible inactivation of the enkephelin receptors, but this inactivation was prevented by the addition of underivatized enkephalin.

Carney *et al.* (1979) have reported the construction of a photoactive derivative of the mitogenic protease, thrombin. N-(4-azido-2-nitrophenyl)-2-diaminoethane (NAPEDE) was conjugated to ^{125}I-labeled human thrombin by carbodiimide-mediated condensation. The resulting NAPEDE−^{125}I-thrombin retained the ability to specifically bind to cultured mouse embryo fibroblasts which respond mitogenic-ally to thrombin. The addition of the NAPEDE group to thrombin by this procedure did not significantly alter the affinity of ^{125}I-thrombin for its receptor.

Table 4.2 Photoreactive derivatives of peptide ligands

Ligand	Cross-linker used	Group derivatized	Reference	Binding affinity for receptors
Concanavalin A	Methyl-4-azidobenzoimidate	8 Amino groups per molecule of tetrameric concanavalin A. Sites of insertion not determined.	Ji, 1977	Specific binding to carbohydrates, but binding kinetics not determined.
Murine EGF	Methyl[3-(p-azidophenyl)-dithio] propionimidate	α-Amino group at the NH_2-terminus.	Das et al., 1977	No loss of binding activity.
Murine EGF	N-(4-azido-2-nitrophenyl)-ethylenediamine	Carboxyl groups. Sites of insertion not determined.	Das and Fox, 1978	No loss of binding activity.
Murine EGF	N-succinimidyl-6-(4′-azido-2′-nitrophenylamino) hexanoate	α-Amino group at the NH_2-terminus.	Hock et al., 1979	Not determined.
Insulin	4-azidobenzoxy-N-hydroxy-succinimide	α-Amino group of B29 lysine	Yip et al., 1980	Slightly lower then that of underivatized insulin.
Insulin	4-Fluoro-3-nitrophenylazide	Amino groups. Sites of insertion not determined.	Jacobs et al., 1979	No loss of binding activity.

(continued on the next page)

Biological activity	Radiolabeling of receptor	Cross-linked complex formed as a linear function of specific binding of the derivatized ligand
Not determined	Apparently labels the band 3 protein of erythrocyte membranes	Not determined
No loss of biological	Specific labeling of a single 3T3 cell surface protein (M_r = 190 000)	Established
Not determined	Specific labeling of a single 3T3 cell surface protein (M_r = 190 000)	Not determined
Not determined	Specific labeling of 2 placenta membrane proteins (M_r = 180 000 and 160 000)	Not determined
67% of that of underivatized insulin	Specific labeling of 2 proteins (M_r = 130 000 and 90 000) in liver membrane	Not determined
Not determined	Specific labeling of an M_r = 135 000 band in both rat liver and human placenta membrane	Established

(continued on the next page)

Table 4.2 Photoreactive derivatives of peptide ligands (*continued*)

Ligand	Cross-linker used	Group derivatized	Reference	Binding affinity for receptors
D-Ala^2-Leu^5 Enkephalin	4-Fluoro-3-nitrophenylazide	ϵ-Amino group of a Lys residue coupled to the C-terminus of the ligand.	Hazum *et al.*, 1979	Binding affinity of the same order as the native ligand.
D-Ala^2-Met^5 Enkephalin	N-(4-azido-2-nitrophenyl)-ethylenediamine and tyrosine	The C-terminal carboxyl group.	Lee *et al.*, 1979	Binding affinity of the same order as the native ligand.
α—Thrombin	N-(4-azido-2-nitrophenyl)-ethylenediamine	Carboxyl groups. Sites of insertion not determined.	Carney *et al.*, 1979	No loss of binding activity.

Table 4.2 Photoreactive derivatives of peptide ligands (*continued*)

Biological activity	Radiolabeling of receptor	Cross-linked complex formed as a linear function of specific binding of the derivitized ligand
Not determined	Not determined	Not determined
Not determined	A specific-linked complex of M_r = 80 500 formed with mouse fibroblasts	Established
Not determined	Not determined	Not determined

4.4 SPECIFIC LABELING OF LIGAND-BINDING RECEPTORS WITH PHOTOREACTIVE LIGANDS

This section describes the use of photoreactive derivatives of ligands to identify receptor proteins. Particular emphasis has been placed on studies performed in this laboratory. In the examples that follow, the photoreactive ligands were characterized, to be certain that specific binding activity was retained after ligand derivatization. It is imperative that a photoreactive ligand meets this test if the results of the labeling experiments are to be interpreted in terms of ligand—receptor complex formation.

4.4.1 Epidermal growth factor receptors

The application of photoaffinity labeling in the identification of receptors and description of their metabolic fates is illustrated by studies on the membrane receptor for epidermal growth factor by Das *et al.* (1977, 1978) and Das and Fox (1978). The strategy employed for labeling EGF receptors in mouse 3T3 cells with two photoderivatives of EGF, PAPDIP-EGF and NAPEDE-EGF, is shown in Fig. 4.1. The procedure involves binding of photoreactive EGF to cells at low temperatures ($4°-20°C$), followed by photolysis of the cell-bound EGF at $4°C$ for 5 min. Low temperatures are preferable for these experiments because EGF receptors are rapidly endocytosed and proteolytically degraded at $37°C$ (Das and Fox, 1978). Photolysis of ^{125}I-labeled PAPDIP-EGF bound specifically at low temperatures to EGF receptors on 3T3 cells resulted in the incorporation of EGF into a single high-molecular weight species. The evidence that this $M_r = 190\ 000$ band represents an EGF-EGF receptor complex is summarized in Table 4.3. Another photoreactive derivative of EGF, NAPEDE-^{125}I-EGF, in which NAPEDE (*N*-(4-azido-2-nitrophenyl)ethylenediamine) was coupled to carboxyl groups in the EGF molecule, was also specifically and irreversibly incorporated into the same $M_r = 190\ 000$ band (Das and Fox, 1978). Treatment of the NAPEDE-EGF-receptor covalent complex with 2-mercaptoethanol did not reduce its molecular weight, making it unlikely that the EGF receptor is a disulfide-linked oligomer.

A time-dependent reduction of radioactivity from the receptor band was observed when cells containing radiolabeled receptor were incubated at $37°C$. The rate of loss of radioactivity was the same irrespective of whether the receptor had been radio-labeled with PAPDIP-EGF or NAPEDE-EGF. The loss of radioactivity from the receptor band was accompanied by appearance of low molecular weight bands of $M_r = 60\ 000$, $47\ 000$ and $37\ 000$. Approximately 90% of the radioactivity lost from the EGF receptor complex was recovered in these low molecular weight species, indicating that they were derived from the $M_r = 190\ 000$ precursor. Upon isopycnic banding of the cellular organelles, these low molecular weight proteins cofractionated with lysosomal enzyme markers enzymes, while the $M_r = 190\ 000$ receptor band co-fractionated with the plasmalemmal marker enzymes. These results show that binding of EGF to its cell surface receptor leads to internalization and lysosomal degradation of EGF-receptor complexes.

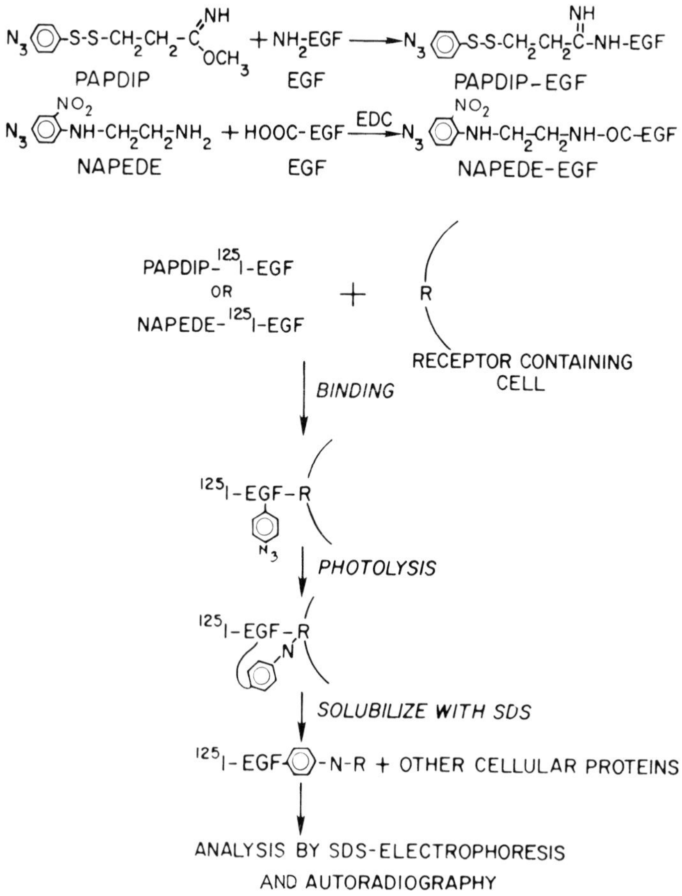

$$N_3 \text{—}\langle O \rangle\text{-S-S-CH}_2\text{CH}_2\text{-C} \overset{\text{NH}}{\underset{\text{OCH}_3}{\diagup}} + \text{NH}_2\text{-EGF} \longrightarrow N_3 \langle O \rangle\text{-S-S-CH}_2\text{CH}_2\overset{\text{NH}}{\underset{}{\overset{\parallel}{\text{C}}}}\text{-NH-EGF}$$

PAPDIP   EGF   PAPDIP-EGF

$$N_3 \overset{\text{NO}_2}{\langle O \rangle}\text{-NH-CH}_2\text{CH}_2\text{NH}_2 + \text{HOOC-EGF} \overset{\text{EDC}}{\longrightarrow} N_3 \overset{\text{NO}_2}{\langle O \rangle}\text{-NH-CH}_2\text{CH}_2\text{NH-OC-EGF}$$

NAPEDE   EGF   NAPEDE-EGF

PAPDIP-^{125}I-EGF
OR   +   R
NAPEDE-^{125}I-EGF

RECEPTOR CONTAINING CELL

| *BINDING*

^{125}I-EGF-R
    N₃

| *PHOTOLYSIS*

^{125}I-EGF-R
    N

| *SOLUBILIZE WITH SDS*

^{125}I-EGF-$\langle O \rangle$-N-R + OTHER CELLULAR PROTEINS

|

ANALYSIS BY SDS-ELECTROPHORESIS
AND AUTORADIOGRAPHY

Fig. 4.1 Procedure for photoaffinity labeling EGF receptors.

In studies performed in this laboratory (Das *et al.,* 1977; Das and Fox, 1978), two photoactive, bifunctional cross-linking reagents were used to covalently affix EGF to its receptors on 3T3 cells. The amino group specific reagent, PAPDIP, was reacted with the terminal amino group of EGF. The reagent, NAPEDE (*N*-(4-azido-2-nitrophenyl)-ethylenediamine) was coupled to an indeterminate number of carboxyl groups on EGF in a reaction driven by a water-soluble carbodiimide. The resulting photoactivatible derivities were then radioiodinated, bound to cells, and photolyzed to generate the highly reactive nitrene group which covalently attached both derivatives to EGF receptors. From Das and Fox (1978).

Table 4.3 Evidence that the cross-linked complex of EGF and a cell surface component is a complex of EGF with its receptors

---

1. EGF complexes to a single membrane protein.

2. No cross-linking is observed with a non-binding, non-responding variant clone of 3T3 cells.

3. The amount of cross-linked complex formed with EGF and the amount of specific binding of ^{125}I-EGF are decreased by the same proportions when unlabeled EGF is added as a competitor.

4. Under a variety of other conditions, e.g. in cells subjected to EGF receptor down regulation, a direct proportionality exists between binding activity and cross-linked complex formation.

After Das and Fox (1978).

The photoaffinity-labeling technique described above also revealed an interesting characteristic of the degradation products of the EGF receptor. When binding, endocytosis and proteolytic processing of EGF receptors proceeded prior to photolysis the low molecular weight receptor-derived products (principle products of M_r = 60 000, 47 000 and 37 000; and a minor product of M_r = 90 000) were covalently labeled with photoreactive ^{125}I-EGF, whereas little or no labeling was observed in the absence of photolysis. The degradation products must therefore retain EGF-binding activity. Studies on the characterization of hormone receptors have been fraught with controversies regarding receptor molecular weights; receptor processing by cellular proteases could explain many of these discrepancies.

Hock *et al.* (1979) undertook the labeling of EGF binding components in human placenta using two different approaches:

(1) gluteraldehyde cross-linking after binding of ^{125}I-EGF to receptor followed by sodium borohydride reduction; and

(2) a photoaffinity labeling approach using ^{125}I-EGF derivatized with *N*-succinimidyl-6-(4'-azido-2'-nitrophenylamino)-hexanoate.

Two labeled components having molecular weights between 150 000 and 200 000 were observed. Because the two methods of radiolabeling the receptor are chemically quite dissimilar, yet yielded identical results, the authors concluded that the risks of misinterpretation inherent in either cross-linking procedure were minimized.

4.4.2 Insulin receptors

The technique of affinity labeling has recently been applied to the study of insulin receptors. Yip *et al.* (1980) used ^{125}I-labeled $N^{\epsilon B29}$-(azidobenzoyl) insulin (^{125}I-MAB-insulin) to identify insulin binding proteins in rat, mouse and guinea-pig liver plasma membranes. Two polypeptides of M_r = 130 000 and 90 000 were specifically cross-linked to photoreactive insulin in membranes from all three species of animals. Using a different photoderivative, 4-azido-2-nitrophenyl-insulin (in which the arylazide group was linked to an unidentified amino group on insulin), Jacobs *et al.* (1979) identified the hepatic and placental insulin receptors as polypeptides of M_r = 135 000 under reducing conditions of polyacrylamide gel electrophoresis. Without reduction, a single high molecular weight band of M_r = 310 000 was observed. Pilch and Czech (1979) used the chemical cross-linker, disuccinimidyl suberate, for identification of insulin receptors in rat adipocytes. Under reducing conditions, the fat cell receptor was identified as a polypeptide of M_r = 130 000. In a more recent study (Pilch and Czech, 1980) these same authors also observed that under non-reducing conditions, the insulin receptor migrated as an M_r = 300 000 species. Taken together, these studies suggest that the plasma membrane insulin binding component is a polypeptide of about M_r = 130 000. However, since the receptor exhibits slower electrophoretic mobility in the absence of reduction, and other labeled proteins are sometimes observed (i.e., the M_r = 90 000 band observed by Yip *et al.*), the insulin–insulin receptor complex is apparently a multisubunit structure.

4.4.3 Thrombin receptors

Carney *et al.* used a photoactive derivitive of thrombin, NAPEDE-^{125}I-thrombin, to identify the receptor for thrombin on mouse embryo cells (1979). A ^{125}I-labeled complex of M_r = 80 500 was formed upon irradiation of the NAPEDE-^{125}I-thrombin bound specifically to these cells. The amount of radioactivity present in this complex represented about 60% of the total amount of NAPEDE-^{125}I-thrombin bound to these cells and increased linearly as a function of binding. No radioactivity was detected in this complex when an 80-fold excess of unlabeled thrombin was included to inhibit the specific binding of NAPEDE-^{125}I-thrombin to its receptors. Since thrombin itself has an M_r = 30 250, the receptor for thrombin must have an M_r of approximately 50 000.

4.5 SPECIFIC LABELING OF LIGAND-BINDING PROTEINS BY DIRECT LINKAGE

Some radiolabeled ligands become directly and irreversibly bound to their receptor proteins without the use of chemical cross-linking reagents. This phenomenon of 'direct linkage' forms the basis for the simplest-available method of affinity labeling of the receptors for two polypeptide ligands: EGF (Linsley *et al.*, 1979; Linsley and Fox, a; Baker *et al.*, 1979) and thrombin (Baker *et al.*, 1979).

4.5.1 Epidermal growth factor

A small portion of the ^{125}I-EGF that binds specifically to intact cells or isolated membranes from a variety of sources becomes directly and irreversibly linked to EGF receptors (Fig. 4.2). Direct linkage complex formation was first demonstrated

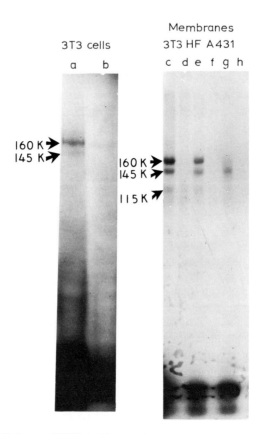

Fig. 4.2 Direct linkage of EGF to its receptors.

Intact 3T3 cells, or isolated membranes from 3T3, HF−15, and A431 cells, were incubated with ^{125}I-EGF in the absence of any photo-activatible reagent. Some samples also contained an excess of unlabeled EGF (lanes b, d, f and h). After a standard binding reaction, cells and membranes were washed, solubilized with SDS, and subjected to SDS-polyacrylamide gel electrophoresis and autoradiography. (From Linsley and Fox, a).

independently in this laboratory on 3T3 cells (Linsley *et al.*, 1979) and on human fibroblasts by Baker *et al.* (Baker *et al*, 1979). A small portion of the ^{125}I-EGF, bound specifically to 3T3 cells, migrated as an M_r = 160 000 protein* having the characteristics of an EGF–EGF receptor complex (Fig. 4.2, lane a). A small amount of an additional complex of M_r = 145 000 was sometimes observed; the formation of both complexes was inhibited when binding proceeded in the presence of unlabeled EGF (lane b). Similar direct linkage complexes are formed with other cell lines. Membranes isolated from cell lines such as 3T3, HF–15 (normal human fibroblasts) and A431 (a human tumor line which displays a high level of EGF receptors) also give rise to several direct linkage complexes; the formation of all is inhibited by excess unlabeled EGF (Fig. 4.2, lanes c–h). Direct linkage complex formation, and ^{125}I-EGF binding exhibit a similar dependence on ^{125}I-EGF concentration. ^{125}I-EGF binding and direct linkage complex formation also have similar susceptibilities to competition by unlabeled EGF.

To determine the role of direct linkage in the mitogenic pathway, we constructed a number of chemically modified derivitives of EGF and assayed them for their abilities to bind and form direct-linkage complexes with the EGF receptor. One derivative, α-amino-biotinyl EGF, bound to EGF receptors on 3T3 cells with the same affinity as native EGF, but was substantially reduced in its ability to form direct linkage complexes. Since this derivative is unchanged in its ability to stimulate the uptake of ^{3}H-Thymidine into DNA, the process of direct linkage does not appear to play an important role in EGF stimulated mitogenesis. Its importance, rather, lies in its usefulness as an affinity-labeling technique for the receptor. The evidence that direct linkage complexes represent ^{125}I-EGF attached to its receptor is summarized in Table 4.4.

The discovery of direct linkage affords an obvious technical advantage since photoaffinity probes are more cumbersome to use for the labeling of EGF receptors. The percentage of specifically bound ^{125}I-EGF involved in direct-linkage complexes varied in different experiments, but was usually less than 2% of the total radioactivity bound at all ^{125}I-EGF concentrations. The yield of direct linkage complex formation, relative to ^{125}I-EGF binding, was increased when the temperature was raised from 23°C to 37°C and when the pH was raised from 7.4 to 8.5. Direct linkage complex formation is slow relative to ^{125}I-EGF binding and does not reach a maximum even after incubations for 48 hour.

Since direct linkage complexes are formed in low yield, we were concerned that a small subclass of EGF receptors, possibly not a representative one, participated in complex formation. This could limit the usefulness of direct-linkage complex formation in affinity-labeling studies of receptor function. A test of this possibility is described in Table 4.5. If membranes are first incubated with unlabeled EGF, a small reactive subclass of EGF receptors might engage in the formation of unlabeled

---

* The EGF–EGF receptor complexes identified by photoaffinity labeling and direct linkage appear identical. The lower apparent molecular weight of the direct linkage complex (M_r = 160 000 as compared to an M_r = 190 000 for the photoaffinity-labeled complex) is caused by a modification in the gel system we employed in later experiments.

Table 4.4 Evidence that direct linkage complexes represent ^{125}I-EGF attached to its receptors

1.	The direct-linkage complex formed with intact 3T3 cells is precipitated from detergent solution by an antiserum to EGF (Linsley *et al.*, 1979).
2.	The direct-linkage complexes and the EGF–EGF receptor complex produced by photoaffinity labeling have identical electrophoretic behavior (Das *et al.*, 1977; Linsley *et al.*, 1979).
3.	The complexes have biological and chemical properties expected of EGF–EGF receptor complexes (Das *et al.*, 1977; Linsley *et al.*, 1979).
4.	Cell surface proteins having properties identical to those of the direct linkage complexes can be identified by surface-specific iodination of EGF-responsive cells (Wrann and Fox, 1979).
5.	After A431 membranes have been solubilized with solutions containing Triton X-100, direct-linkage complex-forming activity co-fractionates with ^{125}I-EGF binding activity during gel chromotography and velocity sedimentation (Linsley and Fox, a).

Table 4.5 Effects of prior incubation of A431 membranes with unlabeled EGF or [127]I-EGF on direct linkage complex formation

First incubation (0–24 h)	Second incubation (24–48 h)	
Addition	^{125}I-EGF bound (fmol)	^{125}I-EGF in direct linkage complexes (fmol)
None	161	3.8
Unlabeled EGF	151	3.1
[127]I-EGF	182	3.3

Duplicate aliquots of membrane protein were incubated with the indicated addition for 24 h at 37°C. At this point, identically-prepared samples specifically bound 223 fmol of ^{125}I-EGF, of which 6.7 fmol were determined to be in direct-linkage complexes. The membranes were then washed and incubated with ^{125}I-EGF for a second 24 h period at 37°C. Specific binding was measured for each sample and the amount of ^{125}I-EGF in direct-linkage complexes was determined after electrophoresis. From Linsley and Fox, a.

direct linkage complexes, thereby reducing the number of receptors capable of forming labeled direct-linkage complexes during a second incubation with ^{125}I-EGF. Alternatively, if the EGF receptors which engage in direct linkage-complex formation are members of a larger class which is more representative of the total population, the small amount of unlabeled complex formed after even an extended incubation period would not noticeably alter the amount of direct linkage complexes formed with ^{125}I-EGF. The data presented in Table 4.5 support the second alternative: prior incubation of membranes with unlabeled EGF did not significantly reduce the amount of direct linkage complexes formed. The lack of inhibition by unlabeled EGF is not due to an inability of non-iodinated EGF to form direct-linkage complexes, since ^{127}I-EGF (EGF labeled with non-radioactive Na^{127}I) was also ineffective in reducing direct linkage complex formation by ^{125}I-EGF. Since the fraction of EGF receptors capable of engaging in complex formation does not appear to be restricted, the use of direct linkage complex formation as a technique for the affinity labeling of EGF receptors for biological studies is justified.

The mechanism of direct linkage is unclear. The direct linkage complexes, once formed, are stable and have resisted methods of disruption which might provide insight into the mechanism of their formation. The complexes are not disrupted by boiling in SDS solutions containing 2-mercaptoethanol. Nor are they disrupted by extremes of pH or treatment with high concentrations of a strong nuclophile, hydroxylamine. The formation process is not inhibited by inhibitors of the trans-glutaminase reaction such as EDTA or exogenously added amines. Regardless of the mechanism by which these complexes are formed, the process of direct linkage is an exquisitely simple and powerful technique for affinity labeling of EGF receptors.

4.5.2 Thrombin

Thrombin also forms direct linkage complexes with a receptor protein on normal human fibroblasts (Baker *et al.*, 1979). This thrombin receptor appears to be a $M_r = 38\,000$ protein, termed protease-nexin (Baker *et al.*, submitted, which is found not only on the cell surface, but also in the cytoplasm. It is also released into the culture medium. The thrombin–protease nexin direct linkage complex differs somewhat from the EGF–EGF receptor direct linkage complexes; it is formed in much higher yield (about 30%) and is disrupted by treatment at alkaline pH or with hydroxylamine, conditions which do not disrupt EGF–EGF receptor direct linkage complexes. Thrombin blocked at its catalytic site does not form this complex. The thrombin–protease nexin complex has properties which closely resemble a reported linkage between thrombin and antithrombin III, a prominent inhibitor of thrombin in serum. These results suggest, by analogy with the role of antithrombin III in regulating the proteolytic activity of thrombin in serum, that protease-nexin regulates the activity of serine proteases at or near the cell surface.

4.6 COVALENT EGF–EGF RECEPTOR COMPLEXES IN THE MAPPING OF EGF RECEPTOR FUNCTIONAL SITES

EGF receptors, affinity labeled by the direct linkage process, have been exploited to make several observations on the disposition of EGF receptor functional sites.
During membrane-isolation procedures the EGF receptor is processed

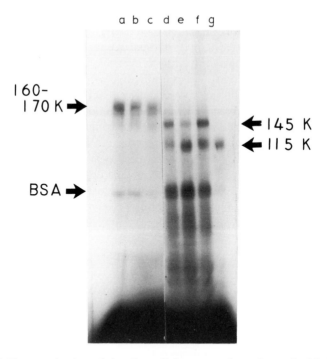

Fig. 4.3 Characterization of the direct linkage complexes formed with A431
cells and membranes.

    The direct linkage complexes formed with intact or treated A431 cells
(lanes a−c) were compared with those formed with similarly treated isolated
membranes (lanes d−g). Some samples were treated with trypsin before
(lanes b, e and f) or after (lanes c and g) ^{125}I-EGF addition. For the sample
run on lane f, soybean trypsin inhibitor was mixed with trypsin prior to
addition to membranes. (From Linsley and Fox, a).

proteolytically. The direct linkage complex formed with intact A431 cells, which
display extraordinary levels of EGF binding, migrated as a diffuse bond of
$M_r = 160-170\ 000$ (Fig. 4.3, lane a) and was relatively insensitive to trypsin
(lane b). Trypsin had little effect also on the capacity of EGF receptors to engage
in direct complex formation when added prior to EGF (lane c). After A431 cells
were scraped from their substratum at the begining of the membrane preparative
procedure, they no longer gave rise to the diffuse $M_r = 160-170\ 000$ direct-linkage
complex but instead displayed increased amounts of the $M_r = 160\ 000$, $145\ 000$
and $115\ 000$ direct-linkage complexes arising as the result of endogenous proteolytic
activity(ies).

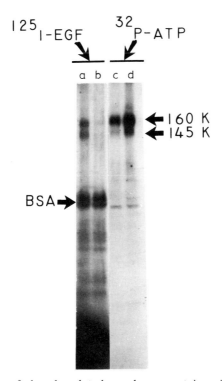

Fig. 4.4 Comparison of phosphorylated membrane proteins with the direct-linkage complexes.

The A431 membrane proteins labeled by the rapid EGF, stimulated-phosphorylation reaction described by Carpenter *et al.* (1978) were compared with those which link directly with EGF. ^{125}I-EGF was added to the samples run in lanes (a) and (b); the sample run in lane (b) also contained excess unlabeled EGF. The samples in lanes (c) and (d) were incubated with (lane c) or without (lane d) unlabeled EGF before the addition of ATP. (From Linsley and Fox, a).

The M_r = 145 000 component visible with isolated membranes (lane d; see also Fig. 4.2) was formed by an activity termed 'scraping protease' which was activated when cells were scraped from their substratum. The M_r = 115 000 component of the direct-linkage complex formed with membranes also can be produced by the action of endogenous protease. The M_r = 160 000 and 145 000 complexes are sensitive to exogenous trypsin, added either before (lane e) of after (lane g) direct-linkage complex formation; the product of this reaction is also an M_r = 115 000 complex.

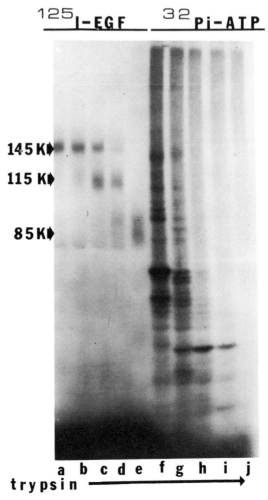

Fig. 4.5 Trypsin sensitivity of phosphorylated membrane proteins.
   A431 membranes were first treated with ^{125}I-EGF (lanes a–e), to form direct-linkage complexes, or with unlabeled EGF and γ-^{32}Pi-ATP (lanes f–i), to phosphorylate the EGF receptors. Samples were then treated with increasing concentrations of TPCK-treated trypsin; the samples run in lanes (a) and (f) received no trypsin. (From Linsley and Fox, a).

We have also utilized direct-linkage to demonstrate that two fragments of the EGF receptor co-migrate during gel electrophoresis with proteins phosphorylated in the rapid, EGF-stimulated reaction described by Carpenter *et al.* (1978). The A431 membrane proteins phosphorylated specifically in response to EGF addition

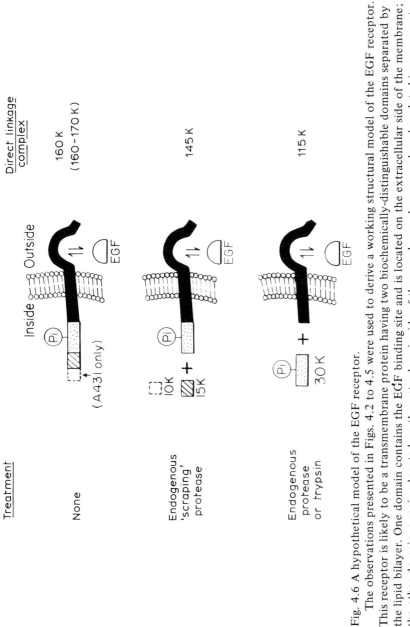

Fig. 4.6 A hypothetical model of the EGF receptor.

The observations presented in Figs. 4.2 to 4.5 were used to derive a working structural model of the EGF receptor. This receptor is likely to be a transmembrane protein having two biochemically-distinguishable domains separated by the lipid bilayer. One domain contains the EGF binding site and is located on the extracellular side of the membrane; the other domain, a region located on the cytoplasmic side of the membrane, becomes phosphorylated in response to EGF. (From Linsley and Fox, a).

migrated during gel electrophoresis with the M_r = 160 000 and 145 000 fragments of the direct-linkage complex (Fig. 4.4). Slight differences in mobility between the fragments of the direct-linkage complex and the phosphorylated proteins were sometimes observed; these differences can be accounted for by the additional 6000 daltons contributed to the complexes by EGF. While co-migration is not a rigorous criterion for protein identity, the fact that two related fragments of the EGF–EGF receptor direct-linkage complex each co-migrate with a protein that is phosphorylated provides strong evidence that the EGF receptor is itself the major protein phosphorylated in the EGF stimulated reaction. We can infer from the data in Fig. 4.3 that the site of phosphorylation on the receptor molecule, like the EGF binding site, is not removed by the 'scraping protease'. Trypsin treatment of ^{32}Pi-labeled A431 membranes under conditions which lead to the production of the M_r = 115 000 fragment of the direct-linkage complex (Fig. 4.5) completely removes the ^{32}Pi incorporated into the M_r = 145 000 fragment (Fig. 4.5). This observation suggests that the site of phosphorylation on the receptor molecule might be located on the M_r = 30 000 portion(s) liberated by trypsin treatment. Another, but less likely, interpretation of this observation is that trypsin treatment of EGF receptors renders labile a phosphate-receptor linkage on the 115 000 dalton fragment. The recovery of the radioactivity released by trypsin from the M_r = 145 000 band as a ^{32}Pi-labeled M_r = 30 000 fragment is required to validate the first interpretation. After trypsin treatment, we did observe that a ^{32}Pi-labeled species of approximately this size accumulated, concominant with the loss of radioactivity in the M_r = 145 000 band (Fig. 4.5). However, since the amount of radioactivity in several other ^{32}Pi-labeled bands also decreased in response to trypsin treatment, we cannot unequivocally determine the origin of this low molecular weight ^{32}Pi-labeled fragment. We are currently repeating the experiment described in Fig. 4.5 with a partially purified ^{32}Pi-labeled EGF receptor preparation.

These observations are summarized in the working structural model of the EGF receptor presented in Fig. 4.6. In this model the cleavage sites for both the 'scraping' protease and trypsin are placed on the cytoplasmic rather than the extracellular side of the membrane, because the EGF receptor in intact cells is relatively insensitive to exogenously added 'scraping' protease and trypsin. Broken cell preparations containing 'scraping' protease activity are ineffective in cleaving the M_r = 160–170 000 direct-linkage complex formed on other, intact cell monolayers. The M_r = 145 000 component of the direct-linkage complex present in isolated membranes is totally converted to the M_r = 115 000 component by concentrations of trypsin which are without effect on the M_r = 160–170 000 complex present in intact cells (Fig. 4.2 and data not shown).

REFERENCES

Baker, J.B., Low, D.A., Simmer, R.L. and Cunningham, D.D., submitted.
Baker, J.B., Simmer, R.L., Glenn, K.C. and Cunningham, D.D. (1979), *Nature*, **278**, 743–745.

Carney, D.H., Glenn, K.C., Cunningham, D.D., Das, M., Fox, C.F. and Fenton, J.W. (1979), *J. biol. Chem.*, **254**, 6244–6247.

Carpenter, G., King, L., Jr. and Cohen, S. (1978), *Nature*, **276**, 409–410.

Darfler, F.J. and Marinetti, G.V. (1977), *Biochem. biophys. res. Commun.*, **79**, 1–7.

Das, M. and Fox, C.F. (1978), *Proc. natn. Acad. Sci., U.S.A.*, **75**, 2644–2648.

Das, M. and Fox, C.F. (1979), *Ann. Rev. biophys. Bioeng.*, **8**, 165–193.

Das, M., Fox, C.F. (1978), In: *Transmembrane Signalling* (Bitensky, M., Collier, R.J., Steiner, D.F. and Fox, C.F., eds), Alan R. Liss, Inc., New York.

Das, M., Miyakawa, T. and Fox, C.F. (1978), In: *Cell Surface Carbohydrates and Biological Recognition* (Marchesi, V.T., Ginsburg, V., Robbins, D.C. and Fox, C.F., eds.), pp. 674–656, Alan R. Liss Inc., New York.

Das, M., Miyakawa, T., Fox, C.F., Pruss, R.M., Aharonov, A. and Herschman, H.R. (1977), *Proc. natn. Acad. Sci., U.S.A.*, **74**, 2790–2794.

Fleet, G.W.J., Knowles, J.R. and Porter, R.R. (1972), *Biochem. J.* **128**, 499–508.

Hazum, E., Chang, K.J., Schechter, Y., Wildinson, S. and Cuatrecasas, P. (1979), *Biochem. biophys. res. Commun.*, **88**, 841–846.

Hixson, S.H. and Hixson, S.S. (1975), *Biochemistry*, **14**, 4251–4254.

Hock, R.A., Nexo, E. and Hollenberg, M.D. (1979), *Nature*, **277**, 403–405.

Huang, C.K., Richards, F.M. (1977), *J. biol. Chem.*, **252**, 5514–5521.

Jacobs, S., Hazum, E., Schechter, Y. and Cuatrecasas, P. (1979), *Proc. natn. Acad. Sci., U.S.A.*, **76**, 4918–4921.

Ji, T.H. (1977), *J. biol. Chem.*, **252**, 1566–1570.

Kiehm, D.J. and Ji, T.H. (1977), *J. biol. Chem.*, **252**, 8524–8531.

Knowles, J.R. (1972), *Acc. chem. Res.*, **5**, 155–160.

Lee, T.T., Williams, R.E. and Fox, C.F. (1979), *J. biol. Chem.*, **254**, 11787–11790.

Lewis, R.V., Roberts, M.F., Dennis, E.A. and Allison, W.S. (1977), *Biochemistry*, **16**, 5650–5654.

Linsley, P.S., Blifeld, C., Wrann, M. and Fox, C.F. (1979), *Nature*, **278**, 745–748.

Linsley, P.S. and Fox, C.F., (a), submitted.

Linsley, P.S. and Fox, C.F., (b), submitted.

Mikkelson, R.B. and Wallach, D.F.H. (1976), *J. biol. Chem.*, **251**, 7413–7416.

Peters, K. and Richards, F.M. (1977), *Ann. rev. Biochem.*, **46**, 523–551.

Pilch, P.E. and Czech, M.P. (1979), *J. biol. Chem.*, **254**, 3375–3381.

Pilch, P.E. and Czech, M.P. (1980), *J. biol. Chem.*, **255**, 1722–1731.

Pinner, A. (1883), *Ber*, **16**, 352–363.

Reiser, A., Willets, F.W., Terry, G.C., Williams, V. and Marley, R. (1968), *Trans. Faraday Soc.*, **64**, 3265–3275.

Traut, R.R., Bollen, A., Sun, T.-T., Hershey, J.W.B., Sundberg, J. and Pierce, L.R. (1973), *Biochemistry*, **12**, 3266–3273.

Wang, K. and Richards, F.M. (1974), *Israel Journal of Chemistry*, **12**, 375–389.

Wrann, M. and Fox, C.F. (1979), *J. biol. Chem.*, **254**, 8083–8086.

Yip, C.C., Yeung, C.W.T. and Moule, M.L. (1980), *Biochemistry*, **19**, 70–76.

5 Immunochemical Identification and Characterization of Membrane Receptors

O. J. BJERRUM, J. RAMLAU, E. BOCK and
T. C. BØG-HANSEN

Acknowledgements

Mrs A. Christophersen and Mrs K. Norup Jensen are thanked for their skilful technical assistance.

We are grateful to Dr T. Plesner for the radiolabelling of the anti-WGA antibody preparation and to Pharmacia Fine Chemicals for delivering of WGA coupled to Sepharose 4B. Financial support has been provided by Kong Christians X's Fond, Bloddonorernes Forskningsfond (O.J.B.) and P. Carl Petersens Fond and Danish Medical Research Council (J.R. and E.B.).

Membrane Receptors: Methods for Purification and Characterization
(*Receptors and Recognition,* Series B, Volume 11)
Edited by S. Jacobs and P. Cuatrecasas
Published in 1981 by Chapman and Hall, 11 New Fetter Lane, London EC4P 4EE
© Chapman and Hall

5.1 INTRODUCTION

Most methods available for the characterization of a given protein require the protein to be present in pure form and in suitable amounts. Such conditions are difficult to fulfil for most membrane proteins because of the problems involved in their isolation. Therefore we have taken interest in some methods that are not limited by these requirements. In contrast to most other immunological methods, quantitative immunoelectrophoretic methods are sensitive, simple to carry out and allow distinction and characterization of a multiple of different antigens at the same time. Thus it is unnecessary to purify individual membrane proteins or to produce specific antibodies, because crude protein mixtures can be examined directly by means of polyspecific antisera. The immunochemical recognition of a receptor combined with characterization by physico-chemical techniques then allows the receptor to be characterized in molecular terms.

This chapter will describe how it is possible, after solubilization with detergents, to identify membrane receptors by their specific interactions in quantitative immunoelectrophoresis. Furthermore it will be demonstrated how characterization at the molecular level of a receptor identified in this way can be performed without previous purification. The use of antibodies for monitoring fractionation procedures will be illustrated, whereas their exploitation as tools in preparative procedures such as immunosorbent techniques will not be described. To illustrate the techniques, two receptor systems will be shown: the nicotinic acetylcholine receptor from the *Torpedo marmorata* electric organ (Mattson *et al.*, 1979) and the MN glycoprotein (glycophorin) from human erythrocyte membranes. These are chosen as examples of a lectin receptor (Wiedmer, 1974; Adair and Kornfeld, 1974; Tanner and Anstee, 1976) and a virus (Howe *et al.*, 1971) receptor. The present knowledge of the two receptors will not be reviewed. Our intention is that the readers will relate the examples to their own receptor systems and apply the techniques to their own projects. Receptor molecules are abundant in the chosen model systems, (e.g. for the MN glycoprotein 5×10^5 copies per cell (Steck, 1974) allowing direct application of immunoelectrophoretic methods for the analysis. However, for analysis of receptors present in smaller amounts it is necessary to use various amplification steps for increasing the sensitivity of the immunoelectrophoretic technique (see Section 5.2.1).

The general strategy for receptor characterization and isolation by means of quantitative immunoelectrophoretic methods will be described. It involves the following steps:

(a) production of a polyspecific antibody;
(b) definition of a reference pattern by crossed immunoelectrophoresis of

solubilized membrane protein, including the receptor;

(c) identification of the receptor in the precipitation pattern;

(d) molecular characterization of the receptor by various biochemical and physico-chemical analyses linked to the immunoelectrophoretic analysis;

(e) testing in analytical scale the possible fractionation methods by means of immunoelectrophoretic 'table-top' techniques;

(f) final isolation on a large scale, with subsequent molecular characterization regarding aspects not covered by the previously used techniques.

5.2 GENERAL REMARKS ABOUT QUANTITATIVE IMMUNOELECTROPHORESIS

5.2.1 Techniques

The quantitative immunoelectrophoresis methods are based upon electrophoresis in antibody-containing gel as devised by Laurell (1965) and Clarke and Freeman (196? and differ in this way from immunoelectrophoretic analysis as described by Grabar and Burtin (1964). There are four main features of quantitative immuno-electrophoresis:

(1) The resolution is based upon specific recognition of individual proteins by their corresponding antibodies, allowing detailed analysis of individual proteins in crude protein mixtures.

(2) The biological activity of the antigen is often retained after immuno-precipitation. This allows further characterization of the proteins, for example with respect to receptor functions.

(3) The area of the immunoprecipitate is proportional to the amount of antigen, and inversely proportional to the antibody concentration, allowing quantification of proteins.

(4) The medium is agarose gel, whose large pores allow analysis of heterogeneous large protein complexes (e.g. membrane proteins), and which furthermore is easy to handle, allowing design of composite experiments.

Three manuals have been published recently giving all the technical details of the present state of the technique (Laurell, 1972; Axelsen *et al.,* 1973; Axelsen, 1975). The literature was reviewed extensively up to 1975 by Verbruggen (1975). Owen and Smyth have covered application in microbiology, (1976) and earlier we have discussed the special condition for the study of membrane proteins by these methods (Bjerrum and Bøg-Hansen, 1976b; Bjerrum, 1977). In the present connection the three most useful methods are crossed immunoelectrophoresis, crossed immuno-electrophoresis with intermediate gel and fused rocket immunoelectrophoresis.

We prefer to perform the crossed immunoelectrophoresis of Clarke and Freeman (1967) instead of that described by Laurell (1972) because the former method is quantitative and easier to perform. The first dimension electrophoresis is performed in an antibody-free gel which is transferred to another supporting glass plate, where an antibody-containing gel is cast in contact with the first dimension gel. Then electrophoresis is performed into the antibody-containing gel, and precipitates are formed overnight (Fig. 5.1). The proteins form a precipitation pattern: with polyspecific antibodies each protein has its characteristic position, shape of precipitate and precipitate morphology, (Fig. 5.1). This reference pattern constitutes a framework for the characterization of all proteins appearing in the pattern.

In crossed immunoelectrophoresis with intermediate gel (Svendsen and Axelsen, 1972; Svendsen, 1973) (Figs. 5.7 and 5.8), an additional gel is interposed between the first dimension and the second dimension gel. Such a gel may contain either antibodies, antigens or ligands (e.g. hormones/toxins or substrates/inhibitors) for specific interactions with the proteins under investigation or the antibodies in the reference gel (Figs 5.2 and 5.7). Fused rocket immunoelectrophoresis (Svendsen, 1973; Harboe, 1979) is a variant of rocket immunoelectrophoresis (electroimmunoassay) (Laurell, 1972). Here a preliminary diffusion of samples applied in adjacent wells takes place in an antibody-free gel, prior to electrophoresis into the antibody-containing gel. This method is primarily used for analysis of fractions from a fractionation experiment to give the elution profiles of individual proteins (Section 5.6 and Figs. 5.16 and 5.17).

The sensitivity of quantitative immunoelectrophoretic techniques permits as little as 10–100 ng of protein to be detected with the conventional staining technique (Coomassie Brilliant Blue R), the sensitivity depending on the antigen–antibody system examined (Laurell, 1972; Weeke, 1973a). However, an increase in sensitivity of 10–60 fold can be obtained by employing enzymatic or autoradiographic methods (see for example Bjerrum, 1977). Thus the membrane antigen may be labelled with [125]I (Nørgaard-Pedersen, 1973; Raftell and Blomberg, 1974; Nørgaard-Pedersen and Axelsen, 1976), with ^{14}C or [35]S (Westin, 1976; Vestergaard and Grauballe, 1975) or even with [3]H (Norén and Sjöström, 1979), or alternatively the immune plate may be incubated after electrophoresis with specific ^{125}I-labelled antibodies (Kindmark and Thorell, 1972; Plesner, 1978). A more general approach is incubation with enzyme-labelled (Brogren et al., 1976; Kjaervig and Inglid, 1981), or ^{125}I-labelled (Rowe, 1969) swine or sheep antibodies against rabbit γ-globulin (the most commonly employed species). Also electrophoretic application, at pH 5, of labelled swine antibodies against rabbit γ-globulins can be employed (Ramlau and Bjerrum, 1977).

Most of these enhancing systems would allow detection of even smaller amounts of proteins, but are restricted by the precipitation limit of the antigen–antibody system. Thus, well defined precipitates are not formed using antigen amounts less than approximately 0.1–1 ng (Ramlau and Bjerrum, 1977, Brogren and Bøg-Hansen, 1975). In cases where a receptor is present in these or smaller amounts, a prefractionation must be preformed before the analysis can take place, in order to remove ballast proteins and to obtain a higher concentration, since no more than

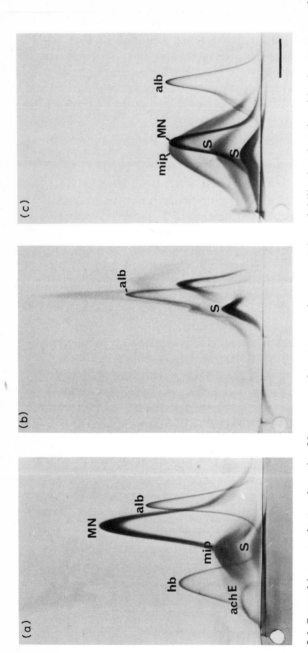

Fig. 5.1 Crossed immunoelectrophoresis of human erythrocyte membrane proteins solubilized and analysed in the presence of various detergents: (a) Triton X-100 (solubilization-conc gel-conc⁻¹): 1.0:0.5% (v:v), (b) SDS 1.0:0.1% (w:v) and (c) Empigen BB (1.0:0.2% (w:v)). All detergent concentrations are above their respective CMC*, except for the SDS gel concentration. Human albumin, 0.5 μg, has been added to each preparation. In the precipitation pattern the following proteins have been identified (Bjerrum and Bøg-Hansen, 1976a): albumin (alb), spectrin (s), MN glycoprotein (MN), protein complex containing the major 'intrinsic' protein (mip) Band III protein, acetylcholinesterase (achE) and haemoglobin (hb). Note the number of precipitates and the different migration velocity relative to albumin of the individual proteins in various detergents. All gels are 1% agarose containing 0.038 M Tris and 0.1 M glycine, pH 8.7. The second dimension gel contains antibodies to human erythrocyte membrane proteins (Dakopatts, Copenhagen) in a concentration of 4.5 μl cm⁻². First dimension electrophoresis was performed at 10V cm⁻¹ and second dimension immunoelectrophoresis at 3V cm⁻¹ for 18 h. The plates were stained for esterases (1-naphthylacetate, Fast Red TR) and proteins (Coomassie brilliant blue). The bar represents 1 cm.

* CMC: critical micellar concentration.

50–200 μl of material can be applied in crossed immunoelectrophoresis. Greater volumes can be applied using rocket immunoelectrophoresis under certain conditions (Krøll, 1976b). A rough impression of the receptor material needed for immuno-electrophoretic analysis can be obtained from the precipitation pattern for human erythrocyte membrane proteins shown in Fig. 5.1(a). About 20 μg of membrane protein was applied, corresponding to approximately 4×10^7 cells (assuming 5.7×10^{-10} mg of protein per haemoglobin-free ghost (Steck, 1974). The major precipitates (spectrin, major intrinsic protein, Band III protein and MN glycoprotein; see Fig. 5.1(a)) correspond to proteins present to the extent of about 5×10^5 molecules per cell. Acetylcholinesterase with approximately 6×10^3 molecules per cell could be detected by its enzymatic activity but not by protein staining (Bjerrum, 1977). The above mentioned figures suggest that, in order to obtain membrane protein precipitates, between 10^7 and cells with more than $10^4 - 10^5$ molecules per cell of each protein are necessary. These considerations may be changed if the receptor exists as a complex with other antigenic proteins which are more frequently represented in the membrane (Bjerrum, 1977; Norrild *et al.*, 1977).

5.2.2 Antibodies

An absolute requirement for the use of these methods is the availability of antibodies against the protein or protein mixture under study. Usually rabbits are chosen for production of antibodies since they provide antibodies in appropriate amounts (Bjerrum, 1977). The animals are immunized for more than a year with monthly bleedings because the antibody titer increases during the course of immunization, especially against the minor components in the immunogen mixture.

When starting a production of antibodies against a new complex protein system (protein mixture) it is preferable to use several rabbits in order to eliminate the individual differences in the antibody response, and also to increase the amount of antibody produced (an analysis requires large amounts of antibody). An 'ordinary' crossed immunoelectrophoretic experiment usually requires between 0.3 and 3 ml of antiserum, and it is recommended that a large pool of antiserum ($>$ 500 ml) is collected before the animals are killed, if needed.

For the immunization, crude membranes, solubilized membrane protein or fractionated protein may be used as immunogens in doses of 50–500 μg per injection. With pure immunogens as little as 25 μg protein per kg body weight has been used with good results. The rabbits are injected and bled regularly (every two to four weeks) according to the scheme of Harboe and Ingild (1973).

The first pool of purified antibodies is used for the subsequent standardization of the procedures for solubilization and purification of the protein preparation. The final standard preparation then defines the reference pattern in crossed immuno-electrophoresis. Having established the reference pattern, it is easy – by comparison – to study the individual proteins and their distribution and occurrence in the succeeding antibody preparations. Axelsen *et al.*, thoroughly discussed the problems connected with comparison of antigens and antisera (1973a).

When the main object is to produce specific anti-receptor antibodies, for example for use in histological studies on localization or in preparative experiments for separation of the receptor by immunoadsorption, the quantitative immunoelectrophoresis methods can provide a shortcut (Shivers and James, 1967). Production of specific antibodies can often be obtained by cutting out the immunoprecipitated individual antigen directly from the agarose gel after an immunoelectrophoresis experiment. The agarose-enclosed precipitate is used directly as immunogen for immunization (Vestergaard, 1975; Krøll, 1976a; Berzins *et al.,* 1977; Section 5.5.3).

Fractionation of the antiserum to obtain the immunoglobulin fraction permits the use of a higher antibody concentration in the gels, and the background staining of the immunoplates is reduced at the same time. The other proteins which are present in unfractionated antiserum may cause artefacts by interacting with the antigen under examination (Bjerrum and Bøg-Hansen, 1976b; Gardner and Rosenberg, 1969). For example, in the case of sodium dodecylsulphate-solubilized membrane antigens, the presence of the detergent can give rise to false precipitates by interaction with serum lipoproteins (Green *et al.,* 1975, Carey *et al.,* 1975). The lipoproteins of th antiserum may also influence certain amphiphilic proteins (Bøg-Hansen, unpublished).

5.3 SOLUBILIZATION OF MEMBRANE RECEPTORS
FOR IMMUNOCHEMICAL ANALYSIS

5.3.1 Principles of solubilization

At present the solubilization of membrane proteins for the investigation of their receptor properties is best achieved with non-ionic detergents, since the membrane proteins are then maintained at conditions which mimic those in the membrane (Tanford and Reynolds, 1976; Helenius and Simons, 1975; see Chapter 2). Addition of surplus detergent to a biological membrane results in a re-arrangement of the amphiphilic molecules present in the system. With non-ionic detergent, solubilization of the membrane principally appears to result from a replacement of lipid molecules by detergent molecules. The membrane lipids, as well as the amphiphilic 'integral', or 'intrinsic' membrane proteins (Singer and Nicholson, 1972), become incorporated into the dominating, water-soluble detergent micelles, resulting in membrane disintegration. Protein—protein interactions are usually not affected during this process, and proteins are seldom denatured. Detergent binding is confined to the apolar surfaces of membrane proteins; hydrophilic 'peripheral' membrane proteins do not significantly bind such mild detergents. Thus a membrane sample solubilized with non-ionic detergent comprises a mixture of delipidated, 'integral' membrane proteins to which detergent is bound, 'peripheral' membrane proteins to which detergent is not bound, and mixed lipid-detergent micelles (Helenius and Simons, 1975; Tanford and Reynolds, 1976).

Ionic detergents often denature proteins because they bind both to hydrophobic and hydrophilic domains of a protein molecule in a cooperative way (Tanford and Reynolds, 1976).

From the above considerations it follows that the presence of non-ionic detergents in a protein solution need not affect the hydrophilic molecular regions of proteins. The receptor sites, being primarily confined to such hydrophilic surfaces, remain available for interaction with a specific ligand. This is the basis for studying receptor—ligand binding and for performing immunochemical analyses of proteins in detergent solution.

5.3.2 Practical solubilization procedure

Although procedures for solubilization of membranes with non-ionic detergents may require modifications from system to system, observation of the following principles generally leads to satisfactory protein solubilization (Bjerrum, 1977; Helenius and Simons, (1975). The protein concentration should not exceed 3—4 mg ml^{-1}, and a detergent: protein weight ratio of 3:1 or more is desirable. Solubilization of peripheral membrane proteins may be enhanced by low ionic strength ($I < 0.05$), pH changes (pH 5—7 and 8—12) and removal of divalent metal ions by addition of chelators (e.g. EDTA). To minimize formation of disulfide bridges and undesirable enzymatic effects, inhibitors (e.g. iodoacetamide and protease inhibitors as for example aprotinin, phenyl methyl-sulfonyl fluoride and pepstatin) may be added and the solubilization performed at 0—4°C. Sonication of aggregated membrane preparations for short periods (2—10 s at 20 000 Hz) may help to promote solubilization after addition of detergent. As a definition, solubilized membrane material is recovered as supernatant after ultracentrifugation at 100 000 g for 60 min. However, in many instances, centrifugation at 40 000 g for 20—30 minutes will be sufficient to remove aggregates which otherwise appear as smears at the application site in the immunelectrophoretic analysis.

In troublesome cases, it is profitable to test several detergents where the hydrophile-lipophile balance (Egan *et al.*, 1978) and the critical micellar concentration (CMC) of the detergents are important key factors (Helenius and Simons, 1975; Slinde and Flatmark, 1976). Furthermore, the solubilization procedure with the most promising detergents should be tested for such parameters as time, detergent—protein ratio, pH and ionic strength (Tanford and Reynolds, 1976).

5.3.3 Detergent-immunoelectrophoresis

Removal of the detergent from the amphiphilic membrane proteins results in aggregation, for which reason it is essential to incorporate detergent in the agarose gel during the immunoelectrophoretic analysis. Otherwise the procedures follow those for conventional immunoelectrophoresis (Axelsen *et al.*, 1973) as outlined by Bjerrum and Bøg-Hansen, 1976b. Incorporation of different non-ionic detergents in the gel may change the precipitation pattern (Bjerrum and Bhakdi, 1981). Selective solubilization (Johansson *et al.*, 1975; Liljas *et al.*, 1974; Yu and Steck, 1975) or varying dissociative effects of the detergents on protein complexes may be responsible for changes in the number of precipitates. Variations in electrophoretic migration may be due to differences in the sizes of the different detergent micelles which are bound to the proteins (Helenius and Simons, 1975), thereby reducing the average charge density of the proteins. The appearance of asymmetric precipitates or even double peaks indicates molecular heterogeneity of the solubilized antigen. As monomerisation of membrane proteins by solubilization is seldom complete, membrane precipitates often appear with some trailing in the precipitation pattern. Differences in the relative area below the precipitates reflect, besides the amount of protein applied, differences in the solubilization efficiency of the given detergents. Furthermore, variations in size and migration velocity of the protein–detergent complexes may change the area of a precipitate. The concentration of non-ionic detergent in the gel should be above the CMC (Bjerrum, 1977). 0.1–1% (weight to volume ratio) is generally suitable. Below the CMC, aggregation of 'integral' membrane proteins often occurs, and this collectively gives rise to large, irregular, blurred precipitates which hamper the immunoelectrophoretic analysis (Bjerrum and Bøg-Hansen, 1976a). Furthermore, as a result of the dissociation of the detergent 'micelle' from the 'intrinsic' proteins and their increased average charge density, they have a higher migration velocity. Normally the proteins will be delipidated by the action of the detergent, therefore proteins whose antigenicity or ligand binding are dependent on the presence of lipid may lose their activity. However, the function can be restored by use of special detergents (Maire *et al.*, 1976) or simply by performing the immuno-electrophoretic analysis in detergent-free gels.

Quantitative immunoelectrophoresis of proteins in the presence of ionic detergents is possible (Bjerrum, 1977; Bjerrum and Bhakdi, 1981; Bjerrum *et al.*, 1975) but is hampered by several factors. This is apparent from Fig. 5.1, where human erythrocyte membrane proteins, which have been solubilized and analysed in the presence of SDS (Fig. 5.1(b)), are compared with those solubilized and analyzed with the non-ionic detergent Triton X-100 (Fig. 5.1(a)). Fewer precipitates are seen with SDS, due to denaturation. Furthermore, the allotment of the charge of SDS to the protein molecules minimizes the resolution in the first dimension electrophoresis. In a similar manner, SDS binds to antibody molecules during the second dimension immunoelectrophoresis, causing electrophoretic removal of the immunoglobulins from the agarose gel. On the other hand, lowering of the detergent concentration gives rise to aggregation phenomena as already described for non-ionic detergents (Bjerrum and Bhakdi, 1981). These undesirable effects of ionic detergents can be partially counteracted by the introduction of non-ionic detergents into the gels (Bjerrum *et al.*, 1975; Converse and Papermaster, 1975). Thus a membrane sample can be solubilized with an ionic detergent and then analyzed in gels containing non-ionic detergent (1% or more). The protein-bound ionic detergent is then replaced with non-ionic detergent during the immunoelectrophoresis, with concomitant antigenic renaturing. By this method it is possible to perform SDS polyacrylamide gel electrophoresis for the first dimension, and directly combine it with a second dimension immunoelectrophoresis (Converse and Papermaster, 1975, Chua and Blomberg, 1979). The acetylcholine receptor of *Torpedo california* (Gordon *et al.*, 1977) and the MN glycoprotein of human erythrocytes (Bjerrum and Bhakdi, 1981) have been examined in this way. However, it should be noted that the presence of free SDS in the sample can give rise to artefactual precipitation lines (Gardner and Rosenberg, 1969; Green *et al.*, 1975; Carey *et al.*, 1975).

Neutral zwitterionic detergents carry both positively and negatively charged groups, but their net charge in the pH interval of 3–11 is essentially zero. Two examples are Empigen BB (Allen and Humphries, 1975) and the sulfobetaine series (Goenne and Ernst, 1978). The advantage afforded by these detergents is a high solubilizing efficiency which does not affect the electrophoretic migration of antigens and antibodies. An analysis of human erythrocyte membrane protein in the presence of 1% (weight to volume) Empigen BB is shown in Fig. 5.1(c). However, these detergents are denaturing (e.g. haemoglobin and acetylcholinesterase could not be demonstrated in Fig. 5.1(c)) (Bjerrum and Bhakdi, 1981).

5.4 IDENTIFICATION OF A RECEPTOR

5.4.1 Introduction

The ability of receptors to react with specific ligands can be used for the identification of the receptors by quantitative immunoelectrophoretic methods and their modifications. Here the term receptor will be used in rather a loose way, since no distinction will be made between the physiological ligand, the other agonists, or the antagonists. The aim is to demonstrate affinity, and not efficacy.

A number of methods will be discussed, along with their various advantages, disadvantages and requirements, for example the antibodies against a receptor-containing antigen mixture or the ligand. Furthermore, labelled or immobilized ligand may be utilized in some approaches.

5.4.2 Methods based on interactions occuring in, or detected in, the first dimension electrophoresis

If a ligand, possessing an electrophoretic mobility significantly different from that of a receptor, is bound to the receptor by incubation prior to electrophoresis, a change in migration of the receptor may be observed. Such an alteration is obtained with human erythrocyte membrane glycoproteins by binding to wheat germ agglutinin (WGA) as shown in Fig. 5.2. It is recommended, especially for the detection of

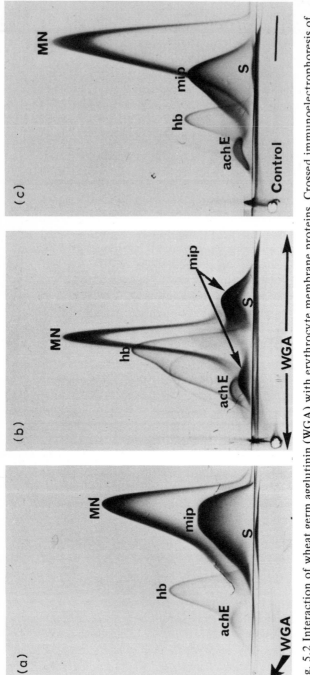

Fig. 5.2 Interaction of wheat germ agglutinin (WGA) with erythrocyte membrane proteins. Crossed immunoelectrophoresis of 20 μg Triton X-100 solubilized proteins: (a) WGA (0.5 mg ml⁻¹) added to the solubilized proteins before electrophoresis. (b) WGA (50 μg/cm²) added to the one dimensional gel. (c) Control. Note in (a) the changes in shape of precipitates of the MN glycoprotein (MN) and the major 'intrinsic' protein (mip). In (b) the MN glycoprotein shows a pronounced retardation, and so does part of the mip complex, indicating heterogeneity. That the retardation was caused by specific binding was demonstrated by incorporating 10% (w/v) N-acetylglucosamine to the sample and one dimensional gel, thereby abolishing the observed changes. For unknown reasons the area below the haemoglobin precipitate (hb) is increased in (b). One- and two-dimensional electrophoresis were performed at 10 V cm⁻¹ for 1 h and 2 V cm⁻¹ for 18 h, respectively. The concentration of the anti-erythrocyte membrane antibodies was 5 μl cm⁻². Other descriptions as for Fig. 5.1.

minor changes in migration, that the first dimension is run in parallel with a control to the incubation mixture, and that an internal migration marker is used (Weeke, 1973b). When intact erythrocytes were agglutinated with WGA, washed, lysed, and the membranes washed and subsequently solubilized and analyzed, only minor changes appeared in the precipitation pattern of the MN glycoprotein. The changes are more pronounced when WGA is mixed with the solubilized material (Fig. 5.2(a)). Both tailing, and a reduction of the area below the MN glycoprotein precipitate (MN), are apparent when compared to the control (Fig. 5.2(c)). Also, the major 'intrinsic' membrane protein (mip) shows a change in shape and in migration velocity.

The method does have drawbacks. When intact membranes are incubated with the ligand, the binding between ligand and receptor must have an association constant high enough to preserve the receptor–ligand complex intact throughout solubilization and the first dimension electrophoresis. Also, mobility differences between receptor and ligand are required, and the receptor and ligand should be of comparable size. Thus, no retardation was observed with the nicotinic acetylcholine receptor from the *Torpedo marmorata* electric organ after incubation with the cathodically migrating α-neurotoxin from *Naja naja*. In this case the receptor is approximately 40 times larger (Mattsson *et al.*, 1979) than the ligand, so the number of positive charges conferred to the complex was not enough to effect a change in migration. A related method utilizes ligand incorporated in the first dimension gel (Bøg-Hansen *et al.*, 1975), and this has some advantages. When migrating in ligand-free gel, a receptor–ligand complex with a low association constant will soon dissociate, and no change in migration will be seen. However, when migrating in a gel with a uniform concentration of ligand, the receptor will be saturated with ligand to an extent determined by the concentration of ligand and by the equilibrium constant at the temperature and pH of the gel.

This system is demonstrated in Fig. 5.2(b), where solubilized erythrocyte membrane proteins are analysed with soluble WGA in the first-dimension gel. It is obvious, by comparison with the control (Fig. 5.2 (c)), that this set-up results in a more marked retardation of the MN glycoprotein than does the incubation experiment (Fig. 5.2(a)). This modification might be better suited for demonstrating receptor–ligand pairs with low affinity, as in the case of the MN glycoprotein. Reduced migration shows up not only in the first dimension, but may also result in a smaller area below the immunoprecipitate of the retarded antigen.

A second advantage is that dissociation constants can be calculated from experiments with ligand in the first dimension (Bøg-Hansen and Takeo, 1980). The dissociation constant, valid for the temperature and pH used for electrophoresis, can be determined if ligand is present in surplus, in a known concentration, thoughout the first dimension gel. Furthermore, the migration of both the receptor and the ligand may be calculated. (Bøg-Hansen and Takeo, 1980; Takeo and Kabat, 1978; Bøg-Hansen, 1979).

The valency of the reactants can also be ascertained from such experiments. If both ligand and receptor are multivalent with respect to each other, they will not

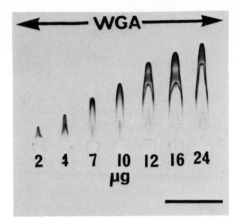

Fig. 5.3 Lectin affinityelectrophoresis of purified MN glycoprotein (glycophorin) of human erythrocyte membranes (Marchesi, 1972). The applied amount of protein (μg) is stated on the figure. The gel contains WGA in a concentration of 50 μg cm^{-2}. Electrophoresis was performed at 2 V cm^{-1} for 12 h in Tris/barbital buffer (pH 8.7, $I = 0.02$).

only bind in a double-molecule complex but will form an affinity precipitate (Bøg-Hansen *et al.*, 1975; Bøg-Hansen *et al.*, 1977; Bøg-Hansen and Takeo, 1980). Fig. 5.3 shows how the purified MN glycoprotein (Marchesi, 1972) can form such affinity precipitates when it is subjected to electrophoresis in a gel containing soluble WGA; if it is directly solubilized from erythrocyte membranes with non-ionic detergent, it does not precipitate (Fig. 5.2(b)). This might be due to aggregation of the purified protein, whereby polyvalency with respect to the carbohydrate groups was attained. I-blood group substances have been analysed in the same way (Owen and Salton, 1976). For a further discussion of affinity-electrophoresis see Bøg-Hansen *et al.* (1977) and Bøg-Hansen (1979, 1980).

In cases where the migration of the receptor–ligand complex is unchanged, compared to the migration of the receptor alone, another modification can be used: the ligand is covalently linked to a suitable matrix and the immobilized ligand is incorporated in the first dimension gel. The precipitate of the receptor will totally or partially disappear from the reference pattern if the receptor binds to the immobilized ligand with a sufficiently high affinity. If the affinity is lower, the precipitate will appear at a retarded position (see later). A system with immobilized toxin is shown in Fig. 5.4(a), where the affinity matrix incorporated in the first dimension gel consists of α-*Naja* toxin coupled to CNBr-activated Sepharose beads (approximately 2 mg toxin ml^{-1} sedimented gel). The antigen is a crude membrane preparation from the electric organ of *Torpedo marmorata*, solubilized in non-ionic detergent. Fig. 5.4(b) represents the control. The precipitate marked with arrow in Fig. 5.4(b) has disappeared in Fig. 5.4(a), thus identifying it as the receptor.

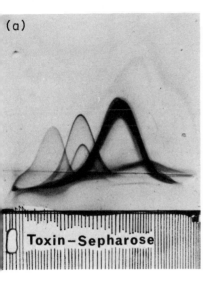

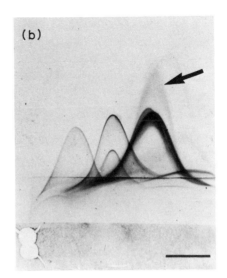

Fig. 5.4 Crossed immunoelectrophoresis of *Torpedo* electroplax membrane proteins with a first dimension gel containing immobilized α-*Naja* toxin.

The antigen is identical in both (a) and (b), and consists of approximately 30 μg of membrane proteins, solubilized in Triton X-100. The antibody against crude electroplax membranes is present at a concentration of 5 μl cm^{-2} in (a) and (b). The first dimension gel of (a) (hatched) contains 40 μl cm^{-2} of a 50% suspension of α-*Naja* toxin coupled to CNBr-activated Sepharose 4B, while (b) is the control, containing the same amount of Sepharose 4B. First dimension electrophoresis is performed at 10 V cm^{-1} for 60 min, second dimension electrophoresis for 18 h at 2 V cm^{-1}. All gels are 1% agarose containing Tris/barbital buffer (pH = 8.6, I = 0.02) and 0.2% Triton X-100. The arrow in (b) points out the precipitate corresponding to the receptor which is absent in (a).

The receptor–ligand complex can also be immobilized after its formation by incorporating antibody against the ligand in the first dimension gel. This system is shown in Fig. 5.5, where *Torpedo* electric organ membrane proteins are analyzed with the standard antibody in the second dimension. In Fig. 5.5(a), the membranes have been incubated with α-*Naja* toxin before solubilization with non-ionic detergent, while Fig. 5.5(b) represents the unincubated control. The first dimension contains antibody towards the toxin. It can be seen that the receptor has disappeared completely from the reference gel in Fig. 5.5(a), and an immunoprecipitate, consisting of the precipitated receptor–ligand complex, can be seen just in front of the application well. In Fig. 5.5(c) and (d), toxin-incubated and control membranes are analysed with 'normal' first dimensions. The receptor has not changed its migration by the binding of toxin, while the precipitate has become more blurred,

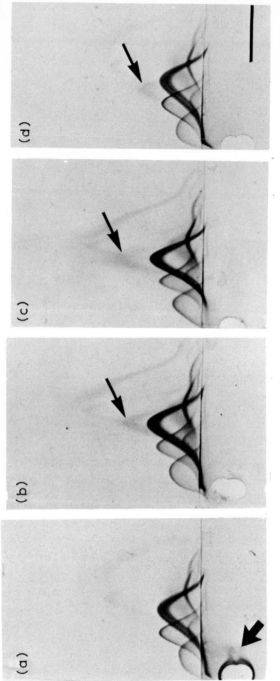

Fig. 5.5 Crossed immunoelectrophoresis of *Torpedo* electroplax membrane proteins incubated with toxin and with antitoxin in the first dimension gel.

The antigens are identical in (a) and (c), consisting of 15 μg of Triton X-100 solubilized *Torpedo* electroplax membrane incubated with 0.13 μg α-*Naja* toxin. The same amount of solubilized membrane protein is used in the controls (b) and (d), but without toxin. The antibody against crude electroplax membranes is present in a concentration of 5 μl cm^{-2} in all plates. In (a) and (b), the first dimension gel contains 5 μl cm^{-2} of a monospecific antibody against α-*Naja* toxin. The first dimension electrophoresis was performed at 10 V cm^{-1} for 35 min. Thin arrows in (b), (c) and (d) point to the receptor, which is absent in the reference gel in (a), but present in the receptor/ligand complex precipitated in front of the well in (a) (fat arrow). Conditions otherwise as for Fig. 5.4. The bar represents 1 cm.

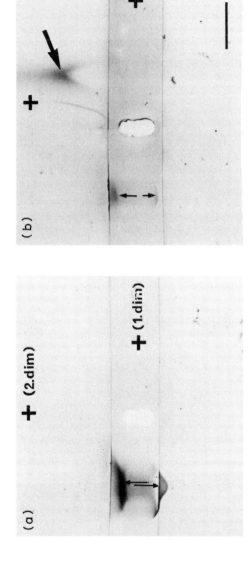

Fig. 5.6 Crossed immunoelectrophoresis of α-*Naja* toxin incubated with solubilized *Torpedo* electroplax membrane proteins. The antigen in (a) is 0.3 μg of purified α-*Naja* toxin mixed with approximately 100 μg of solubilized *Torpedo* electroplax membrane proteins, while (b) is the control, containing the same amount of neurotoxin, but no membrane protein. The first dimension electrophoresis (1 dim) was performed at 10 V cm^{-1} for 35 min. The second dimension gel (2 dim; on both sides of the first dimension separation gel) contains 5 μl cm^{-2} of a monospecific antibody against the α-*Naja* toxin. Thin arrows indicate the free cathodically migrating neurotoxin in both (a) and (b), while the fat arrow points to an anodically migrating complex of toxin and something else, presumably the acetylcholine receptor. Note that the mobility of the complex is similar to the mobility of the receptor. Conditions otherwise as for Fig. 5.4. The bar represents 1 cm.

possibly due to the masking of antigenic determinants (see later discussion). A large surplus of ligand in the incubation should be avoided, as it might absorb the antibody in the first dimension gel, thus preventing precipitation of the complex. Such a risk is absent in the present system, as the excess α-toxin migrates cathodically. Lastly it should be mentioned that the systems described can be turned around, i.e. they can be used to detect changes in the behaviour of the ligand, interpreting these changes as being induced by interaction with the receptor. A possible advantage of this application is that no antibody against the putative receptor is required. This may be valuable in screening several species or tissues for antigenically different receptors with identical ligands.

An experiment utilizing this approach is seen in Fig. 5.6, which shows crossed immunoelectrophoresis of purified α-*Naja* toxin. In Fig. 5.6(a), the toxin is mixed with solubilized membrane proteins from the electric organ of *Torpedo*. Fig. 5.6(b) shows the control with pure α-toxin as antigen. It is obvious that something in the solubilized membrane material has conferred a different migration to the normally cathodically migrating toxin. The method might be used in titrating the ligand-binding capacity of a given preparation: the endpoint is reached when no ligand is present at its original position.

5.4.3 Detection of ligand—receptor interaction with intermediate gel technique

The same principles as were described above for antibody or ligand in the first dimension can be used in the intermediate gel technique (Svendsen and Axelsen, 1972; Axelsen, 1973). The antibody against the ligand is incorporated in the intermediate gel and an incubation mixture containing the putative receptor—ligand complex is analyzed in this set-up. The receptor will be identified by the reduced height of the corresponding precipitate, as the precipitation of the receptor—ligand complex will have begun already in the intermediate gel by the reaction of the complex with the anti-ligand antibody. The method presents an advantage compared to the method with anti-ligand antibody in the first dimension. Thus, if a surplus of free ligand is present, it will often be separated from the receptor—ligand complex by the first dimension electrophoresis.

This set-up is shown in Fig. 5.7. Figure 5.7(a) shows the pattern obtained by analyzing a mixture of α-toxin and solubilized electroplax membranes with anti-toxin in the intermediate gel, and with standard antibody against the membrane antigens in the reference gel. Fig. 5.7(b) shows the pattern of a control without α-toxin, analyzed with the same antibodies. Another control utilizes both incubation mixture and control mixture, analyzed without anti-toxin in the intermediate gel (not shown). It can be seen that one precipitate has been reduced in height, thus identifying it as the receptor. No retardation in the first dimension is obtained. An advantage of the method over the similar set-up with anti-toxin in the first dimension is that the low voltage gradient applied in second dimension might favor precipitation of ligands with low-avidity antibodies. Immobilized ligand can also be used in an

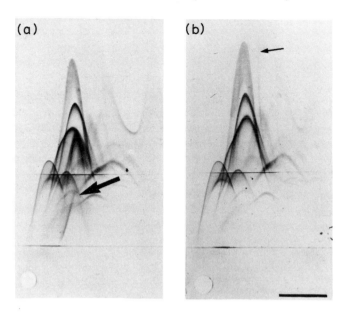

Fig. 5.7 Crossed immunoelectrophoresis with intermediate gel (with antibody against α-*Naja* toxin) of electroplax membranes incubated with α-*Naja* toxin.

The antigen in (a) is 30 μg of solubilized electroplax membranes, which has been incubated with α-*Naja* toxin at a membrane protein: toxin ratio of 60:1 (w:w). The antigen in (b) is incubated without toxin. The reference gel contains 4 μl cm^{-2} of an antibody preparation against electroplax membranes and the intermediate gel contains 5 μl cm^{-2} of a monospecific antibody against α-*Naja* toxin. Identical set-up in (a) and (b). The intermediate gel of (a) contains a precipitate (arrows) not seen in (b), which in return has a precipitate at the same electrophoretic position in the upper gel. The latter must represent the receptor, which is precipitated as complex with the toxin by the anti-toxin in the intermediate gel. Conditions otherwise as for Fig. 5.4. The bar represents 1 cm.

intermediate gel, as affino-immunoelectrophoresis (Bøg-Hansen, 1975). Fig. 5.8 shows such an experiment, where Triton X-100-solubilized human erythrocyte membrane proteins were examined in crossed immunoelectrophoresis with an intermediate gel containing WGA-Sepharose 4B. Compared to the control (Fig. 5.8(b)) the PAS 3 glycoprotein (PAS 3) (Bjerrum *et al.*, 1980) has disappeared, while the MN glycoprotein and the complex containing the major intrinsic protein (mip) have been partially retained. When samples containing smaller amounts of protein are analyzed, the MN glycoprotein is found to have been completely removed, while the same fraction of the normal mip complex remains unbound. This shows that the partial binding observed in Fig. 5.8(a) was an overload phenomenon for MN glycoprotein, but for mip it is probably due to heterogeneity in the carbohydrate moiety of the mip-complex.

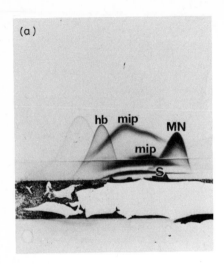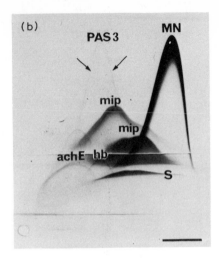

Fig. 5.8 Crossed affino-immunoelectrophoresis of 20 μg Triton X-100-solubilized human erythrocyte membrane proteins.

In (a) the intermediate gel contains wheat germ agglutinin coupled to CNBr-activated Sepharose 4B (50 μg cm^{-2}) and in (b) it is blank. Nearly all the MN glycoprotein (MN) is retained on the WGA-Sepharose and with less load the precipitate completely disappears. The major 'intrinsic' protein (mip) in this preparation is present in two precipitates (Norrild *et al.*, 1977); uncomplexed and complexed mip. In the lower precipitate there is partial binding independent of the load, while the upper precipitate is unaffected. Furthermore, in these membrane preparations a faint precipitate (indicated with arrows) also exhibiting WGA binding properties is observed. It probably corresponds to PAS 3 glycoprotein (Liljas *et al.*, 1974; Bjerrum *et al.*, 1980). In this set up haemoglobin (hb) also shows an increase in area (Fig. 5.2) which cannot be explained. The upper gel contains anti-membrane antibodies (7 μl cm^{-2}). Otherwise the conditions are as for Fig. 5.2.

Generally speaking, the intermediate gel technique is comparable to the first dimension method with immobilized ligand, though the high-voltage gradient applied in the first dimension might be unfavorable for the binding of low-affinity receptors compared to the gradient applied in the second dimension. On the other hand, a weakly-binding receptor might be retarded in the first dimension electrophoresis on the affinity matrix without being bound. This will appear in the reference pattern as a retarded or skew peak but it might not appear as clearly in the approach which utilizes an intermediate gel (Bjerrum, 1978; Bøg-Hansen, 1979).

5.4.4 Detection of bound ligand by other methods

In cases where the ligands are too small to effect a migration change, or where anti-bodies are not available and immobilization is impossible, other methods can be applied. These methods depend on the tagging and subsequent monitoring of the ligands.

Radiolabelling is probably the most suitable method of tagging and will be discussed below. Ligands conjugated with fluorophors may also be used. The choice of radioisotope will be governed by several considerations:

(a) If autoradiography of the immunoplate is to be performed, the energy level of the emitted radiation should be sufficiently high. In our laboratory, good results have been obtained with ^{125}I, ^{35}S, ^{32}P and ^{14}C (Bjerrum and Bøg-Hansen, 1976a; Bjerrum and Bhakdi, 1977; Teichberg *et al.*, 1977). The use of ^{3}H necessitates liquid scintillation counting of the cut-out precipitate as described by Norén and Sjöström (1979) or the incorporation of a scintillating medium in the gel itself (Norén and Sjöström, 1979).

(b) The isotope should be easily incorporated into the ligand.

Both the Chloramine-T method (Hunter, 1971) and the lactoperoxidase method (Marchalomis, 1969; Thorell and Johansson, 1971) are well established for introducing iodine into polypeptides containing aromatic amino acids. The labelling procedure might change the properties of the ligand, either by the oxidation inherent in the method itself or by changing an aromatic amino acid at a possible active site. The possibility of radiation-induced changes after storage of the labelled compound should be borne in mind. The labelled ligand should be tested for biological activity to guard against the above mentioned pitfalls. If the components cannot be labelled with iodine, recourse can be taken to the very wide range of ^{14}C-labelled pharmaco-logically and biologically active substances which are commercially available.

The experimental procedure conceptually consists of two steps:

(a) Binding the ligand to the receptor, and
(b) detecting the bound ligand, and thus the receptor, in the immunoprecipitation pattern.

The ligand can be added to the membranes either before or after solubilization (Mattsson *et al.*, 1979; Blomberg and Berzins, 1975). In some cases (Blomberg and Berzins, 1975) it is possible to incubate the pressed, but still wet, immunoplates in a solution containing the labelled ligand, thus utilizing a sandwich technique.

Another technique has recently been worked out (Kidmark and Thorell, 1972; Plesner, 1978): the solubilized membranes are mixed with unlabelled ligand and analyzed using crossed immunoelectrophoresis with the standard antibody, and this is followed by normal pressing and washing of the plate. The plate is then incubated overnight in a very dilute solution of radio-labelled anti-ligand antibody (50 μg immunoglobulin l^{-1} with a specific activity of 3×10^4 cpm μg^{-1}). After this

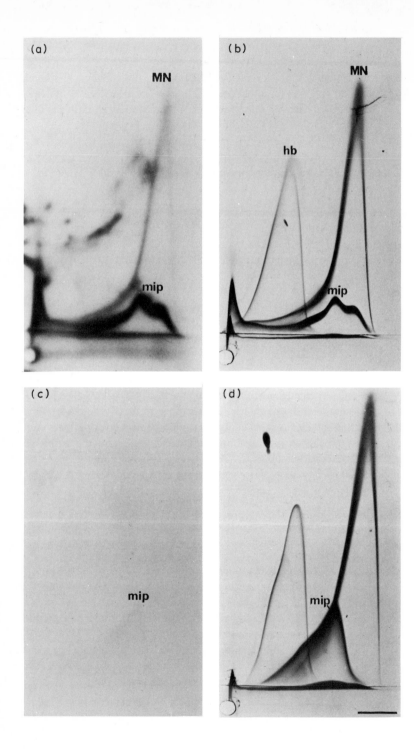

incubation the plate is washed for 24 hours, dried, stained and submitted to auto-radiography. Fig. 5.9 shows such an experiment with WGA as the ligand and human erythrocyte membrane proteins as the source of possible receptors. The solubilized membrane proteins and the ligand were incubated before analysis. After incubation of the plate with ^{125}I-labelled, anti-WGA antibodies, the MN glycoprotein and major 'intrinsic' protein complex were more radioactive than the precipitates of the control plate (compare Figs 5.9(a) and (c)), indicating the presence of WGA in these immunoprecipitates. Non-specific binding of the labelled antibody to other precipitates (as noticed in Fig. 5.9(a)) may reduce the resolving power. The technique can also be used for amplification of the sensitivity of immunoelectrophoresis. When used in rocket immunoelectrophoresis, as little as 0.5–1.0 ng of WGA could be quantified.

Fig. 5.10 shows an experiment where membrane fragments from the *Torpedo* electric organ were incubated with ^{125}I-labelled α-*Naja* toxin and subsequently solubilized in non-ionic detergent and analyzed by crossed immunoelectrophoresis with a polyspecific antibody. After washing and drying, the plate was stained and submitted to autoradiography. The autoradiograph (Fig. 5.10(a)) was then compared to the stained plate (Fig. 5.10(b)). One major toxin-binding antigen (arrow) and several antigens binding lesser amounts of toxin can be seen. The major toxin-binding antigen is the nicotinic acetylcholine receptor (Mattsson *et al.*, 1979). Due to the high sensitivity of the autoradiography, minor amounts of unspecifically bound ligand are often detected. Whether an observed binding is specific or not should therefore be ascertained by trying to abolish binding with a surplus of cold ligand or other agonists or antagonists.

The method can be made quantitative by cutting out the precipitates and assessing the amount of bound radioactivity in the receptor–ligand complex by counting (Christiansen and Krøll, 1973; Section 5.5.3.). This approach makes it

Fig. 5.9 Crossed radioimmunoelectrophoresis with ^{125}I-labelled anti-WGA.
In (a) and (b), 20 μg solubilized human erythrocyte membrane proteins have been mixed with 0.040 μg WGA before electrophoresis. (c) and (d) represent the controls without WGA. (a) and (c) are autoradiographs. (b) and (d) are the same plates after Coomassie Brilliant Blue staining. After electrophoresis the plates were washed, pressed and incubated overnight in a very dilute solution of ^{125}I-labelled anti-WGA (50 μg immunoglobulin l^{-1} with a specific activity of 3×10^4 cpm μg^{-1}) containing 1 g l^{-1} of human albumin. Before the autoradiography, the plates were washed, pressed, dried and finally stained with Coomassie Brilliant Blue (Plesner, 1978). Otherwise the experimental conditions and other designations were as for the experiment of Fig. 5.1 . Exposure time for the autoradioagraphs was 18 h at $90°C$ using a 'Lightning plus' intensifying screen (Dupont) (Lasky and Mills, 1977). Note the labelling of the MN glycoprotein and the mip precipitate in (a) but also the fainter unspecific labelling of the mip precipitate in (c), in spite of intensive absorption of the anti-WGA antibody with a surplus of human erythrocyte ghosts. Figure designations as for Fig. 5.2. The bar represents 1 cm.

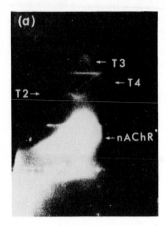 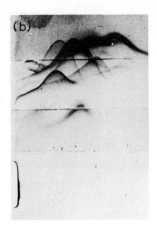

Fig. 5.10 Crossed immunoelectrophoresis of *Torpedo* electroplax membranes incubated with ^{125}I-labelled α-*Naja* toxin. (a) Autoradiograph. (b) The same plate after Coomassie brilliant blue staining. The antigen is a total homogenate of electroplax, incubated with ^{125}I-labelled α-*Naja*-toxin (specific activity approximately 2.5 mCi mg^{-1} protein) at protein to toxin ratio of 250:1 prior to solubilization. The intermediate gel contains a monospecific antibody against the receptor, while the upper gel contains the polyspecific antibody against electroplax membranes. The major part of the radioactive label is situated in the precipitate of the receptor (nAChR), while minor amounts can be seen in other precipitates (T2, T3 and T4). (Reproduced with the kind permission of Journal of Neurochemistry (Mattsson *et al.*, 1979).

possible to calculate a ratio of bound ligand to unit antigen (i.e. receptor), as the area of a precipitate is proportional to the amount of antigen. This determination would be of interest in fractionation work, in order to monitor the intactness of the receptor, since the ratio gives an impression of the ligand binding capacity relative to the antigenicity. This is analogous to the homospecific activity of enzymes, i.e. the ratio of the *in vitro* enzymatic activity to the immunochemically determined amount of protein (Rush *et al.*, 1974).

5.4.5 Discussion

The choice of ligand is important. Though receptors are generally considered to bind ligands, *in vivo* they might very well only do so for a short time before the ligand is removed or enzymatically destroyed. Thus one should not expect to obtain a complex with strong, durable binding by just mixing the receptor and the physiological ligand together. In many cases pharmacological agonists or antagonists with other affinity constants and less liability to enzymatical breakdown might prove advantageous (Heidmann and Changeux, 1978). Solubilization may dissociate the complex (if incubation is performed before solubilization) but non-ionic detergents are generally mild and only interfere with hydrophobic interactions (see Section 5.2).

There are some immunochemical points to consider: antigenic determinants on both receptor and ligand may be masked in the complex. This implies that bound ligand might not be accessible to the relevant antibodies so that no precipitation of the complex can be obtained with the anti-ligand antibody. The same might be the case with the receptor, leading to the disappearance of the receptor from the immunoprecipitation pattern in the reference gel when ligand is added to the receptor. We have observed partial immunochemical identity between the native nicotinic acetylcholine receptor and the α-*Naja* toxin receptor complex. This could be explained either by the blocking of some determinants on the receptor by the toxin, or by loss of determinants due to conformational changes induced by the binding of the ligand.

This consideration also applies to the method of incubating immunoplates in a solution of radioactively-labelled anti-ligand antibody. If the complex of receptor and ligand in the immunoprecipitates is 'covered' with antibodies against the receptor, it may not bind the anti-ligand antibodies. However, since the precipitate is formed in excess of antigen (Kindmark and Thorell, 1972) this situation seems unlikely for ligands of smaller dimensions.

5.5 MOLECULAR CHARACTERIZATION OF A RECEPTOR

5.5.1 Biochemical characterization

Immunoprecipitates can be classified according to whether they contain protein, carbohydrate or lipid antigens.

The protein nature of an antigen can be established by exposure of the solubilized membrane material to proteolytic degradation with enzymes of broad specificity, e.g. pronase, prior to the immunoassay. The disappearance of precipitates then indicates that they are of protein nature (Bjerrum and Bøg-Hansen, 1976a).

Glycoproteins or carbohydrate antigens may be identified in a precipitation pattern by radioactive labelling, by carbohydrate staining, by lectin binding or by changes induced by the action of glycosidases. Direct staining for carbohydrates on the immunoplates can be performed with periodic acid-Schiff or copper-formazan reagents (Howe and Lee, 1969; Niediech, 1978). Since the agarose is slightly stained and since the antibodies contain 3–12% carbohydrate, only those glycoproteins which contain considerably greater proportions of carbohydrates can be distinguished. The MN glycoprotein of human erythrocytes (containing approximately 60% carbohydrate) has been characterized in this way (Howe and Lee, 1969). Sialo-glycoproteins can be identified by their change in migration velocity after neuraminidase treatment. The negative charge of sialic glycoproteins at pH 8.7, and the removal of the sialic acid residues on either intact cells or isolated membranes by neuraminidase

treatment, result in reduced migration velocity in the chosen electrophoresis
system for these glycoproteins after solubilization (Schmidt-Ullrich *et al.*, 1975).
In this way the MN glycoprotein has been shown to contain sialic acid (Bjerrum
and Bøg-Hansen, 1976a).

Demonstration of an interaction between membrane antigens and lectins, 'free'
or immobilized, is another very useful method for identification and characterization
of glycoproteins (Fig. 5.2; Section 5.4; Bøg-Hansen, 1980). Today a large battery
of purified lectins, reacting with nearly all naturally occurring sugar moities, are
commercially available for such identification.

The presence of lipid in immunoprecipitates can be established by staining
techniques which employ the hydrophobic dye Sudan Black (Uriel, 1971).
Detergent-solubilized intrinsic membrane protein does not normally contain lipid
(Bjerrum and Bøg-Hansen, 1976a). However, residual lipid has been shown to be
present in certain multienzyme complexes of cell membranes (Blomberg and
Raftell, 1974). Membrane glycolipids may give rise to immunoprecipitates, but only
in the absence of non-ionic detergent (Niediech, 1978; Argaman and Razin, 1969.

5.5.2 Molecular heterogeneity

Examination of the precipitation pattern in crossed immunoelectrophoresis can
yeild information directly concerning the molecular heterogeneity of the identified
receptor (Owen and Smyth, 1976). Electrophoretic heterogeneity is indicated by
deviation of shape of the precipitate from the normal bellshape which is characteristic
of a homogeneous protein population. The presence of various populations containing the same molecule, or the presence of fragments, may give rise to splitting and/
or doubling of the peaks of the precipitates with concomitant occurrence of spurs
characteristic of partial identity between proteins (Axelsen *et al.*, 1973b; Bjerrum
and Bøg-Hansen, 1976b). Thus the presence of receptor—ligand complexes and
their uncomplexed counterparts can give rise to reactions (Fig. 5.2) which should
not be mistaken for intrinsic heterogeneity of the receptor. Isoelectric focusing and
immunoelectrophoresis in combination can further elucidate such heterogeneity
(Berzins *et al.*, 1976; Schmidt-Ullrich *et al.*, 1977). As shown in Fig. 5.11, erythrocyte
membrane proteins solubilized with non-ionic detergent can be analysed with
crossed immunoelectrofocusing. The major precipitates of Fig. 5.10 are visible.
In contrast to the spectrins, the MN glycoprotein appears to be homogeneous
with a $pI \sim 3$. Optimal separations are generally obtained using gels that contain both
a non-ionic detergent and urea (Bhakdi *et al.*, 1975; O'Farrell, 1975). However,
many membrane proteins may still be separated by isoelectric focusing in gels
containing non-ionic detergent only. This affords the advantage that protein
denaturation by urea is avoided. In our hands, separations obtained in vertically
placed gel rods are the most reproducible. Second dimension immunoelectrophoresis
can easily be performed by placing the longitudinally sliced polyacrylamide gel rod

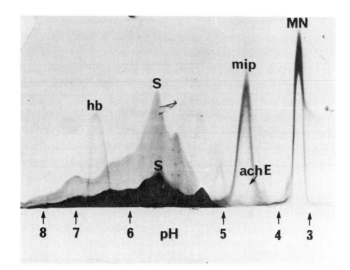

Fig. 5.11 Crossed immunoelectrofocusing of 100 μg Triton X-100 solubilized human erythrocyte membrane proteins (Bjerrum and Bøg-Hansen, 1976b). All the major precipitates are observable but only the spectrins (s) reveal heterogeneity. The electrofocusing was performed in polyacrylamide gel (4% with 2.5% crosslinking) containing 2.5% Triton X-100 (Anode to the right). The second dimension gel contains 0.5% (v/v) Triton X-100 and 15 μl cm^{-2} of anti-erythrocyte membrane antibodies. The immunoelectrophoresis was performed at 2 V cm^{-1} for 16 h. The pH gradient of the first dimension is plotted on the figure. Otherwise the designations and experimental conditions follow those of Fig. 5.1.

(after a wash for 20 min to remove ampholines) on top of an agarose gel positioned as the normal one-dimensional gel (Fig. 5.11; Bjerrum and Bhakdi, 1981). It is necessary to have a layer of agarose gel below the polyacrylamide gel because of the electroendosmotic waterflow in the agarose.

In those cases where the antigenicity of membrane antigens is retained after exposure to sodium dodecylsulphate, heterogeneity in molecular size can be detected by crossed immuno-SDS-polyacrylamide gel electrophoresis. Thus the presence of the monomer and the dimer of the MN glycoprotein has been clearly observed with this method and so has the presence of degradation products of the mip protein (Bjerrum and Bhakdi, 1981). The method will also be advantageous in analysis of cross-linked membrane proteins.

Variation in the degree of glycosylation of glycoproteins can be detected by affinity interaction electrophoresis with 'free' lectin in the first dimension gel (Bøg-Hansen *et al.*, 1977; Bøg-Hansen, 1980).

5.5.3 Amphiphilic properties

Demonstration of detergent-binding properties of a membrane receptor is indicative of its amphiphilic structure and implies its association in the membrane with the phospholipid bilayer (Helenius and Simons, 1975; Singer and Nicholson, 1972; Section 5.3.1). Several techniques are available for performing such a characterization.

In crossed immunoelectrophoresis, upon removal of non-ionic detergent, an increase in electrophoretic mobility and a change of the precipitate morphology is seen for intrinsic membrane proteins, as has been observed for the MN glycoprotein (Bjerrum and Bøg-Hansen, 1976a; Bhakdi *et al.*, 1976). Therefore, such changes are indications of detergent binding for the proteins of the involved precipitates. However, it is by no means a final proof (Bjerrum 1977) because hydrophilic proteins may be affected by the aggregation as well.

Direct demonstration of detergent binding can be carried out by performing the immunoelectrophoresis in the presence of radioactive detergent (Bjerrum and

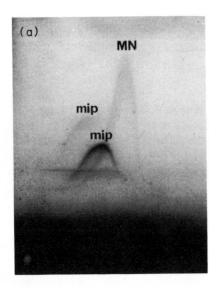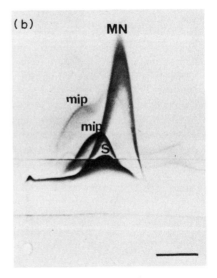

Fig. 5.12 Autoradiographical demonstration of Triton-binding to human erythrocyte membrane proteins after crossed immunoelectrophoresis.
12 µg Triton-solubilized membrane material (Bjerrum and Bøg-Hansen,1976b) is electrophoresed in the presence of ^{125}I-labelled Triton X-100 (Bjerrum and Bhakdi, 1977) (a) Autoradiograph. (b) The same plate after Coomassie Brilliant Blue staining. Only the amphiphilic membrane proteins MN glycoprotein (MN) and major intrinsic proteins (mip) (present in two precipitates) exhibit radioactivity whereas the spectrins (S) and haemoglobin (hb) do not. First and second dimension electrophoresis were run for 45 min at 10 V cm^{-1} and 18 h at 2V cm^{-1}, respectively. Anti-erythrocyte membrane antibodies: 10 µl cm^{-2}. Exposure time for autoradiograph: 10 days. The bar represents 1 cm.

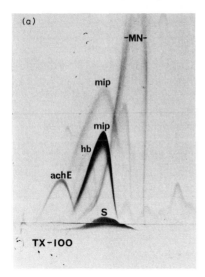

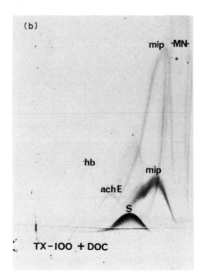

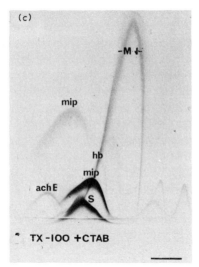

Fig. 5.13 Charge-shift crossed immunoelectrophoresis of 20 μg Triton X-100-solubilized human erythrocyte membrane proteins (Bjerrum and Bøg-Hansen, 1976b.

First dimension electrophoresis was performed in (a) Triton X-100 (Tx), (b) Triton plus deoxycholate (DOC) and (c) Triton plus cetyltrimethyl-ammoniumbromide (CTAB). The precipitates corresponding to haemoglobin (hb) are indicated with an arrow. Marked bidirectional charge-shifts are observed for the 'integral' proteins: MN glycoprotein (MN), major 'intrinsic' protein (mip) (present in two precipitates) and acetylcholinesterase (achE). Anti-erythrocyte membrane antibodies: 7 μl cm^{-2}. The bar represents 1 cm.

Bhakdi, 1977). Precipitates containing amphiphilic proteins will appear more intensively labelled upon autoradiography than the background, due to selective binding of detergent. Fig. 5.12 shows the technique, employing ^{125}I-labelled Triton X-100, applied to human erythrocyte membrane proteins (Bjerrum and Bhakdi, 1977). The MN glycoprotein is radioactively labelled, indicating its amphiphilic nature, whereas the spectrins, classified as 'peripheral' membrane proteins, do not appear in the autoradiography.

Charge-shift crossed immunoelectrophoresis is an indirect way of identifying amphiphilic proteins (Helenius and Simons, 1977). Three different first-dimension electrophoreses are performed in the presence of (a) Triton X-100, (b) Triton X-100 plus the anionic detergent deoxycholate and (c) Triton X-100 plus the cationic detergent *N*-cetyl-*N, N, N*-trimethylammonium bromide. The second dimension gels contain Triton X-100. Because mixed micelles of the two detergents are formed, micellar binding is detected by a change in migration velocity of the respective proteins compared to migration in the presence of Triton X-100 only. In contrast to hydrophilic proteins, only amphiphilic proteins will show a bidirectional shift in migration velocity because of binding of both types of detergent (Bhakdi *et al.*, 1977). Fig. 5.13 shows the results of such analyses of human erythrocyte membrane. Migration distances are related to the hydrophilic marker haemoglobin which, in all three systems, migrates a determined distance e.g. 20 mm. With the electrophoretic conditions employed, the MN glycoprotein shows a bidirectional charge-shift of + 14 mm and −6 mm. Experience with a large range of amphiphilic membrane proteins shows that a bidirectional charge-shift of more than ± 5 mm under these conditions is indicative of detergent-binding properties of the protein (Bhakdi *et al.*, 1977).

Furthermore, the presence of apolar domains in a receptor can be demonstrated using another modification of crossed immunoelectrophoresis: crossed hydrophobic interaction immunoelectrophoresis with phenyl-Sepharose. The amphiphilic proteins of the sample will remain bound to the hydrophobic matrix which is incorporated in the first dimension gel, and they will therefore be absent from the precipitation pattern unless they are liberated from the hydrophobic matrix by displacement with non-ionic detergent incorporated in the second dimension gel (Bjerrum, 1978). The MN glycoprotein is retained on phenyl-Sepharose (Bjerrum, 1978; Fig. 5.14; Section 5.6).

Selective labelling with ^{125}I of intrinsic membrane proteins *in situ* can be performed by means of photosensitive, lipophilic reagents, as for example 5-iodo-naphthyl-1-azide (Bercovici and Gitler, 1978). After incubation with membranes at 37°C, the label, due to its hydrophobic nature, will be present only in the lipid core of the membrane. Subsequent activation with u.v. irradiation binds the label to proteins in the hydrophobic domain. Application of this technique to human erythrocyte membrane proteins gives rise to results corresponding to those in Fig. 5.12. (O. J. Bjerrum and O. Norén, unpublished).

The choice of the immunoelectrophoretic method for the demonstration of amphiphilic properties of a receptor will depend on the antigen–antibody system.

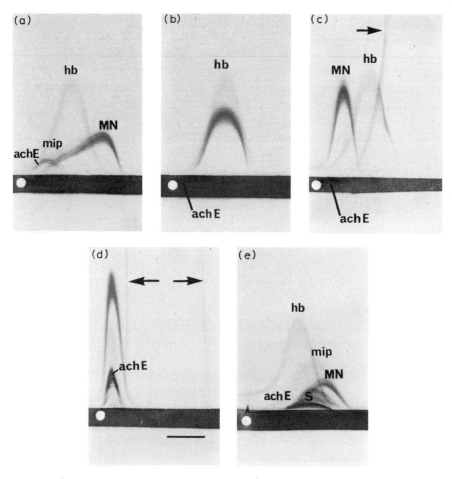

Fig. 5.14 Crossed hydrophobic interaction immunoelectrophoresis (Bjerrum, 1978 of 5 μl Berol EMU-043-solubilized human erythrocyte membrane proteins in the presence of (a) ethyl-agarose, (b) butyl-agarose, (c) hexyl-agarose (d) octyl-agarose, (Miles-Yeda, Rehovot, Israel) in the first dimension electrophoresis. (e) represents the control. Note the increasing retardation of the MN glycoprotein (MN) with increasing alkyl chain length. The first dimension gel does not contain detergent, in contrast to the second dimension gel which contains 1% (v/v) Triton X-100 for the liberation of protein bound to the hydrophobic matrix. Second-dimension gels contained 10 μl cm^{-2} of anti-erythrocyte membrane antibodies. Open arrows indicate an artefactual precipitation line arising from albumin contaminating the first dimension gel. Figure designations and experimental conditions are otherwise as described for Fig. 5.1.

The methods all have their drawbacks, so that demonstration of amphiphilic properties of a receptor is safer if two or more methods are employed. Thus with radioactive detergents only major precipitates can be identified, and with charged detergents the precipitation pattern can be so drastically changed that it is difficult to identify the precipitate in question (Alexander and Kenny, 1978). Furthermore, denaturing can take place in the presence of ionic detergents. With hydrophobic interaction immunoelectrophoresis, detergent is excluded during the interaction with the ligand, and furthermore, specific interaction may take place with the phenyl groups. Iodination with lipophilic labels demand special reagents, and the selectivity has not yet been conclusively demonstrated.

5.5.4 Polypeptide analysis

Individual antigens can be isolated as immunoprecipitates which are obtained by carrying out crossed immunoelectrophoresis on a mixture of labelled antigens. A receptor isolated in this manner can be studied under conditions which usually destroy its antigenicity, making it possible to determine, for example, the polypeptide content and the apparent molecular weight by SDS-polyacrylamide gel electrophoresis (Norrild *et al.*, 1977), isoelectric point and genetic heterogeneity by isoelectric focusing, and all of these by two-dimensional electrophoresis according to O'Farell (Pødenphant *et al.*, 1981).

Labelling of antigens can be performed *in vivo* by incorporation of radioactive precursors, for example an aminoacid or an aminosugar (Vestergaard and Grauballe, 1975). *In vitro* a protein can be labelled by means of ^{125}I by the chloramin T method (Teichberg *et al.*, 1977) or the lactoperoxidase method (Hunter, 1971). If a receptor is a glycoprotein, labelling of the carbohydrate moiety can be performed by the galactose oxidase method (Gahmberg and Hakomori, 1973). However, the label may not necessarily be an isotope, and fluorophors such as fluorescamine may be conjugated to a receptor.

Isolation of a receptor antigen by immunoprecipitation is most conveniently performed by means of crossed immunoelectrophoresis (Vestergaard and Grauballe, 1975). The first dimensional electrophoresis separates the antigen mixture, and the second dimensional electrophoresis reduces unspecific adhesion to the immunoprecipitate, making this method superior to tube precipitation. Furthermore, the method is a must if the antiserum contains antibodies with specificity against several antigens. Since it has been shown that many antibody preparations exhibit proteolytic activity due to the presence of plasmin (Bjerrum *et al.*, 1975), it is advisable to include a protease inhibitor in the antibody preparation and in the subsequent washings to prevent possible degradation of the antigens during the isolation procedure. After immunoelectrophoresis the gel must be washed carefully, and then the immunoprecipitate can be excised and the radioactivity determined. The antigen–antibody complexes are thereafter dissolved by media suitable for the subsequent analytical procedure. Precipitates containing membrane antigens

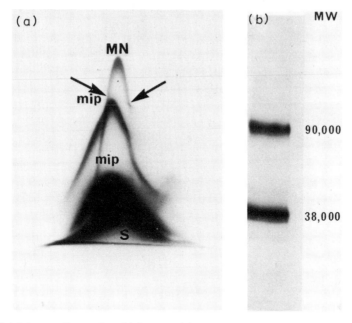

Fig. 5.15 Autoradiographs of (a) crossed immunoelectrophoresis of 20 μg
of ^{125}I-labelled Berol EMU-043-solubilized human erythrocyte membrane
proteins and (b) SDS-polyacrylamide gel electrophoresis (Pødenphant *et al.,*
1981) of isolated MN glycoprotein precipitate cut out of the gel according to
Norrild *et al.* (1977). In (a) the electrophoretic conditions are as for Fig. 5.1.
Note the changed electrophoretic migration of the MN glycoprotein compared
to Fig. 5.1(a). This is due to the use of another nonionic detergent Berol
EMU-043 (Bjerrum and Bøg-Hansen, 1976a). The 'free' part of the precipitates
between the arrows has been cut out for SDS-gel electrophoresis. The two
protein bands PAS 1 and PAS 2 (Fairbanks *et al.,* 1971; Steck, 1974) of the
MN glycoprotein precipitate are clearly seen. Figure designations are as
described for Fig. 5.1.

need both SDS, an S—S reducing agent, and sonication for their complete solubilization (Vestergaard and Grauballe, 1975).

In Fig. 5.15(a) an autoradiograph of the crossed immunoelectrophoresis of
Berol-solubilized ^{125}I-labelled erythrocyte membrane proteins is shown. The MN
glycoprotein precipitate between the arrows is cut out of the gel and subjected to
SDS-polyacrylamide gel electrophoresis and autoradiography (Fig. 5.15(b)). Two
radioactive polypeptide bands of apparent molecular weights of 90 000 and 38 000
are found, corresponding to the dimer and monomer of the MN-glycoprotein also
designated PAS I and PAS II (Steck, 1974). Acethylcholine receptor from
Electrophorus electricus (Heidmann and Changeux, 1978) and lymphocyte
membrane receptors (Walsh *et al.,* 1977), have been cut out and analysed in the
same way.

A prerequisite for the method is that the immunoprecipitates can be cut out free from contamination with other precipitates. Those positioned below other immuno-precipitates may contain small amounts of the antigens of the upper precipitates due to coprecipitation (Brogren and Bøg-Hansen, 1975; Bjerrum and Bøg-Hansen, 1976a). Therefore only the 'free' part (Fig. 5.15) of the precipitates should be cut out and used in the subsequent analysis. To obtain better separation of the immuno-precipitates in crossed immunoelectrophoresis, it is possible to change parameters such as antigen composition, batch of antibodies, buffer composition and type of detergent used (Section 5.3.3.).

If a specific antiserum is available, it should be noted that the described isolation procedure makes it possible to isolate all of the solubilized receptor from a very small volume. This implies possibilities for quantitative determinations of the radioactivity present in the precipitate (Norén and Sjöström, 1979). For example it would be possible to determine the rate of synthesis *in vivo* as well as *in vitro* and the rate of intracellular transport.

5.6 PREDICTION OF ISOLATION PROCEDURES FOR A RECEPTOR

As demonstrated in the preceding sections, the immunoelectrophoretic approach offers possibilities for determination of various molecular parameters of the receptor in the crude mixture obtained after solubilization. On the basis of such characteriz-ation it should be possible to work out fractionation procedures for the subsequent isolation of the receptor. For a direct prediction of the results of preparative scale fractionation, a series of analytical 'table top' techniques has been worked out where the fractionation principle has been combined directly with immunoelectrophoresis. Such methods save both time and material because the fractionation and analysis is preformed simultaneously and all applied material is used for the monitoring.

Crossed affino-immunoelectrophoresis using lectins was the first example of the application of such a compound technique for predicting the outcome of affinity chromatography of serum glycoproteins with concanavalin A (Bøg-Hansen, 1975). Later experiments have confirmed its applicability on solubilized membrane proteins (Schmidt-Ullrich *et al.,* 1975; Berzins and Blomberg, 1975). Fig. 5.8 shows its application on the interaction of erythrocyte membrane proteins with immobilized wheat germ agglutinin. The results correspond to those obtained on a preparative column with the same matrix, as shown in Fig. 5.16. However, parameters such as binding capacity and elution conditions for the column experiment cannot be determined directly from the immunoelectrophoresis experiment.

In a similar way the interaction of a given receptor with various hydrophobic matrices can be tested. Fig. 5.14 shows how solubilized erythrocyte membrane proteins are tested for their interaction with matrices containing alkyl-chains of increasing length. On the basis of the retardation in the first dimension electrophoresis

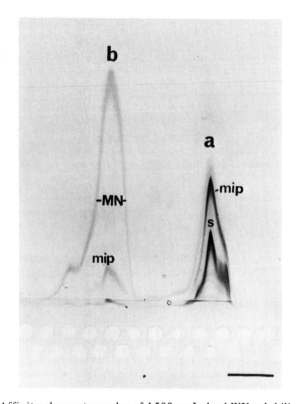

Fig. 5.16 Affinity chromatography of 1500 μg Lubrol WX-solubilized human erythrocyte membrane protein on a 2.5 ml WGA-Sepharose column (Pharmacia Fine Chemical, Uppsala) monitored by fused rocket immuno-electrophoresis. The column is a 5 ml plastic syringe (Bjerrum, 1978). 10 μl aliquots of each fraction (1 ml) are applied as indicated, peak (a) represents the non-retarded proteins eluted with 0.1% (w/v) Lubrol WX in 0.1 M glycine 0.038 M Tris at pH 8.7, and peak (b) represents the proteins eluted with 10% (w/v) of *N*-acetylglucosamine in the above mentioned buffer (arrow indicates the start). The upper gel contains 3 μl cm^{-2} of anti-erythrocyte membrane antibodies. Electrophoresis is performed at 2 V cm^{-1} overnight. Figure designations are as described for Fig. 5.1. The bar represents 1 cm.

it can be predicted that the MN glycoprotein will partially interact with hexyl-agarose but is fully bound to octyl-agarose. This prediction is in accordance with the results obtained with preparative hydrophobic interaction-chromatography (Liljas *et al.*, 1974; Bjerrum, 1978). The results of preparative scale agarose gel electrophoresis (Johansson *et al.*, 1975; Wroblewski, 1977), isoelectric focusing (Berzins *et al.*, 1976), (see Fig. 5.11) and gelfiltration (Nielsen, 1975) can also

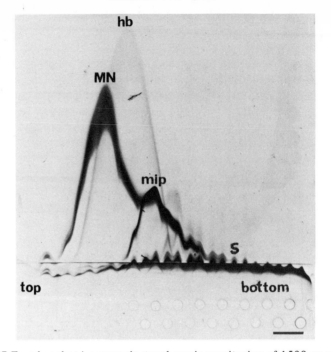

Fig. 5.17 Fused rocket immunoelectrophoresis monitoring of 1500 μg of Triton X-100-solubilized human erythrocyte membrane proteins in 0.116 M phosphate buffer (pH 7.4) and 1% (v/v) Triton X-100 on a 4 cm long Triton X-100-containing (0.5% (w/v)) continous sucrose gradient (4 ml, 10–35% (w/v)) after ultracentrifugation for 40 000 rpm (1.4 x 10^4 g_{av}) for 18 h. 5 μl aliquots of fractions are obtained by dripping out from the bottom as indicated (bottom right). The MN glycoprotein (MN) stays at the top of the gradient. The gel contains 6 μl cm^{-2} of anti-erythrocyte membrane antibodies. Figure designation and experimental conditions are otherwise as described for Fig. 5.16. The bar represents 1 cm.

be predicted by similar combined methods. Ion-exchange matrices cannot directly be incorporated into the immunoelectrophoresis, but in this case other 'table-top' prediction methods have been worked out (Løwenstein *et al.*, 1976).

By means of fused rocket immunoelectrophoresis (Svendsen, 1973), where a large number of fractions are analysed simultaneously, nearly every fractionation procedure may be monitored immunochemically. This technique yields an immunochemical elution profile for each individual protein which appears as a continuous precipitate, indicating the fractions in which the protein is eluted, the relative distribution of the proteins in the fractions and the distribution of contaminating proteins. The method makes it possible to pool fractions containing receptor protein for further fractionation with maximum yield and minimum contamination. Much more information is obtained by this means than by ordinary ultraviolet-absorption monitoring, and the presence of non-ionic detergent, which otherwise can interfere in the latter monitoring technique, has no effect on the results obtained (Bjerrum and Bøg-Hansen, 1976b).

However, it should be born in mind that fused rocket immunoelectrophoresis detects only those components for which antibodies exist and that an immuno-chemically 'pure' fraction may contain other non-immunogenic contaminants. Therefore other techniques such as dodecyl sulphate polyacrylamide gel electro-phoresis must be used as well. With the latter method oligomeric proteins may give rise to confusion. In this case impurities will not be bound by a passage through antibodies incorporated in agarose gel (Bjerrum and Lundahl, 1974) or coupled to a column.

Fig. 5.16 shows how fused rocket immunoelectrophoresis is used for monitoring the outcome of an affinity chromatography of a solubilized erythrocyte membrane proteins on a column containing WGA-Sepharose. By elution with N-acetylglucosamine the MN glycoprotein and a fraction of major 'intrinsic' protein complex eluted from the column in full agreement with the prediction obtained from Figs. 5.2 and 5.8

Other fractionation procedures than chromatographic experiments can be monitored by fused rocket immunoelectrophoresis. Fig. 5.17 shows the distribution of Triton-solubilized erythrocyte membrane proteins on a continuous sucrose gradient (10–35%) after ultracentrifugation where aliquots of the fractions dripped out from the tube have been analysed by fused rocket immunoelectrophoresis.

5.7 CONCLUDING REMARKS

Identification and characterization of receptor molecules among another membrane proteins demand those techniques which are characterized by the following criteria:

(a) A mild solubilization procedure which preserves the receptor fraction and
(b) a high-resolution analytical technique working in non-denaturing conditions.

The methods based upon quantitative immunoelectrophoresis fulfil these require-ments in most instances, as they employ conditions permitting receptors to bind ligands. Most other analytical methods working at conditions which preserve receptor function are of low resolution, but necessarily not of low sensitivity: (e.g. agarose gel electrophoresis, complement fixation, agglutination, test tube immuno-precipitation, simple ligand-binding solid-phage assays and radioimmunoassay). Methods with high resolving power, such as SDS-polyacrylamide gel electrophoresis and two-dimensional isoelectric focusing-SDS-polyacrylamide gel electrophoresis, either work under conditions destroying receptor function, or else the sample preparation results in dissociation of the receptor–ligand complexes. Recourse can be taken to affinity labelling or the use of photo-activated bifunctional reagents for fixing the receptor–ligand complex, but these methods often require organic chemistry at quite a high level.

Furthermore, it is possible to correlate the precipitates of quantitative immuno-electrophoresis with patterns obtained in other high-resolution analytical methods (see Section 5.5). On the other hand the immunoelectrophoresis methods are based on the use of antibodies, which, being a biological product, suffer from drawbacks due to variability with regard to composition, affinity and avidity. It should always be borne in mind that a given receptor might not give rise to an antibody response, either because it constitutes too small a fraction of the total antigenic load presented to the rabbit, or because the receptor is not antigenic. Teleological arguments can be made on both points. It might very well be that key functions are governed by very few receptors, and the more important the function, the fewer the receptors. Thus a putative receptor for botulins toxin would probably only be present in amounts corresponding to the number of molecules present in the minimal lethal dose of the toxin (within an order of magnitude). From the evolutionary viewpoint the receptors might very well be highly-conserved molecules and therefore unrecognizable to the animals as 'non-self'.

Another restriction of the immunoelectrophoretic methods may arise from the possible scarcity of important receptors. The sensitivity of the method is governed not by the sensitivity of the amplification system as such (the sensitivity of autoradiography or the specific activity of the ligand) but by the precipitation limit inherent in the methods. Thus, in the systems studied so far, 'good' precipitates are not formed below antigen amounts of 0.5 ng.

In spite of these limitations, a wide range of membrane receptors are suitable for immunochemical analysis and characterization. Thus, recently, various receptors of chicken erythrocytes (Brogren *et al.*, 1979) lymphocytes (Plesner and Bjerrum, 1979), thrombocytes (Hagen *et al.*, 1979), adipocytes (Pillion *et al.*, 1980) and yeast cells (Gerlach *et al.*, 1979) have been characterized. It is the authors' hope that the many possibilities for receptor identification and characterization outlined in the present chapter will encourage some of the readers to adopt quantitative immunoelectrophoresis as a supplementary tool in their research work. Good luck!

REFERENCES

Adair, W.L. and Kornfeld, S. (1974), *J. biol. Chem.*, **249**, 4696–4704.

Alexander, A.G. and Kenny, G. (1978), *Infection and Immunity*, **20**, 861–863.

Allen, J.C. and Humphries, C. (1975), *FEBS Letters*, **57**, 158–162.

Argaman, M. and Razin, S. (1969), *J. gen. Microbiol.*, **55**, 45–58.

Axelsen, N.H. (1973), *Scand. J. Immunol.*, **2**, Suppl. 1, 71–77.

Axelsen, N.H. (1975), *Quantitative Immunoelectrophoresis, New Developments and Applications*, Universitetsforlaget, Oslo, alias *Scand. J. Immunol.*, **4**, Suppl. 2.

Axelsen, N.H., Bock, E. and Krøll, J. (1973a), *Scand. J. Immunol.*, **2**, Suppl. 1, 101–103.

Axelsen, N.H., Bock, E. and Krøll, J. (1973b), *Scand. J. Immunol.*, **2**, Suppl. 1, 91–94

Axelsen, N.H., Krøll, J. and Weeke, B. (1973), *A Manual of Quantitative Immunoelectrophoresis. Methods and Applications,* Universitetsforlaget, Oslo, alias *Scand. J. Immuno.,* **2,** Suppl. 1.

Bercovici, T. and Gitler, C. (1976), *Biochemistry,* **17,** 1484–1489.

Berzins, K. and Blomberg, F. (1975), *FEBS Letters,* **54,** 139–143.

Berzins, K., Blomberg, F., Kjellgren, M., Smyth, C. and Wadström, T. (1976), *FEBS Letters,* **61,** 77–79.

Berzins, K., Lando, P., Raftell, M. and Blomberg, F. (1977), *Biochim. biophys. Acta,* **48,** 586–593.

Bhakdi, S., Bjerrum, O.J. and Bhakdi-Lehnen, B. (1977), *Biochim. biophys. Acta,* **470,** 35–44.

Bhakdi, S., Bjerrum, O.J. and Knüfermann, H. (1976), *Biochim. biophys. Acta,* **446,** 419–431.

Bhakdi, S., Knüfermann, H. and Wallach, D.F.H. (1975), *Biochim. biophys. Acta,* **394,** 505–557.

Brogren, C.-H., Bisati, S. and Simonsen, M. (1979), *Prot. Biol. Fluids,* **27,** 467–470.

Bjerrum, O.J. (1977), *Biochim. biophys. Acta,* **472,** 135–195.

Bjerrum, O.J. (1978), *Anal. Biochem.,* **90,** 331–348.

Bjerrum, O.J. and Bhakdi, S. (1977), *FEBS Letters,* **81,** 151–156.

Bjerrum, O.J. and Bhakdi, S. (1981), *Scand. J. Immunol.,* Suppl. 1, (in press).

Bjerrum, O.J., Bhakdi, S., Bøg-Hansen, T.C., Knüfermann, H. and Wallach, D.F.H. (1975), *Biochim. biophys. Acta,* **406,** 489–504.

Bjerrum, O.J. and Bøg-Hansen, T.C. (1976a), *Biochim. biophys. Acta,* **455,** 66–89.

Bjerrum, O.J. and Bøg-Hansen, T.C. (1975b), *Scand. J. Immunol.,* **4,** Suppl. 2, 89–99.

Bjerrum, O.J. and Bøg-Hansen, T.C. (1976b), In: *Biochemical Analysis of Membranes* (Maddy, A.H., ed.), 378–426, Chapman and Hall, London.

Bjerrum, O.J., Liljas, L. and Gerlach, J. (1980), unpublished.

Bjerrum, O.J. and Lundahl, P. (1974), *Biochim. biophys. Acta,* **342,** 69–80.

Bjerrum, O.J., Ramlau, J., Clemmesen, I., Ingild, A. and Bøg-Hansen, T.C. (1973), *Scand. J. Immunol.,* **4,** Suppl. 2, 81–88.

Blomberg, F. and Berzins, K. (1975), *Eur. J. Biochem.,* **56,** 319–326.

Blomberg, F. and Raftell, M. (1974), *Eur. J. Biochem.,* **49,** 21–30.

Bøg-Hansen, T.C. (1975), *Anal. Biochem.,* **56,** 480–488.

Bøg-Hansen, T.C. (1979), *Prot. Biol. Fluids,* **27,** 659–664.

Bøg-Hansen, T.C. (1980), *Affinity Chromatography, Les Colloques de l'INSERM* (Egly, J.-M., ed.), Vol. 86, 399–415.

Bøg-Hansen, T.C. (1980b), In: *A Manual of Immunoprecipitation in Gel* (Axelsen, N.H., ed.), *Scand. J. Immunol.,* Suppl., in press.

Bøg-Hansen, T.C., Bjerrum, O.J. and Brogren, C.-H. (1977), *Anal. Biochem.,* **81,** 78–87.

Bøg-Hansen, T.C., Bjerrum, O.J. and Ramlau, J. (1975), *Scand. J. Immunol.,* **4,** 141–147.

Bøg-Hansen, T.C. and Takeo, K. (1980), *J. Electrophoresis,* **1,** 67–71.

Brogren, C.-H. and Bøg-Hansen, T.C. (1975), *Scand. J. Immunol.*, **4**, Suppl. 2, 37–52.

Brogren, C.-H., Peltre, G., Bøg-Hansen, T.C. and Hansen, A. (1976), *Prot. Biol. Fluids*, **24**, 781–786.

Carey, C., Wang, C.-S. and Alaupovic, P. (1975), *Biochim. biophys. Acta*, **401**, 6–14.

Christiansen, AA.H. and Krøll, J. (1973), *Scand. J. Immunol.*, **2**, Suppl. 1, 133–138.

Chua, N.-H. and Blomberg, F. (1979), *J. biol. Chem.*, **254**, 215–223.

Clarke, H.G.M. and Freemann, T. (1967), (1967), *Prot. Biol. Fluids*, **14**, 503–509.

Converse, C.A. and Papermaster, D.S. (1975), *Science*, **189**, 469–472.

Egan, R.W., Jones, M.A. and Leninger, A.L. (1978), *J. biol. Chem.*, **251**, 4442–4447.

Fairbanks, G., Steck, T.L. and Wallach, D.F.H. (1971), *Biochemistry*, **10**, 2606–2617.

Gahmberg, C.G. and Hakomori, S. (1973), *J. Biol. Chem.*, **448**, 4311–4317.

Gardner, E. and Rosenberg, L.T. (1969), *Immunology*, **17**, 71–76.

Gerlach, J.H., Bjerrum, O.J., Rank, G.H. and Bøg-Hansen, T.C. (1979), *Prot. Biol. Fluids*, **27**, 479–482.

Goenne, P. and Ernst, R. (1978), *Anal. Biochem.*, **87**, 28–38.

Gordon, A.S., Davis, C.G., Milfay, D. and Diamond, I. (1977), *Nature*, **267**, 539–540.

Grabar, P. and Burtin, P. (1964), *Immunoelectrophoretic Analysis*, Elsevier, Amsterdam.

Green, J., Dunn, M.J. and Maddy, A.H. (1975), *Biochim. biophys. Acta*, **382**, 457–461.

Hagen, I., Bjerrum, O.J. and Solum, N.O. (1979), *Eur. J. Biochem.*, **99**, 9–22.

Harboe, N. (1979), *Prot. Biol. Fluids*, **27**, 3–10.

Harboe, N. and Ingild, A. (1973), *Scand. J. Immunol.*, **2**, Suppl. 1, 161–164.

Heidmann, T. and Changeux, J.-P. (1978), *Ann. Rev. Biochem.*, **47**, 317–359.

Helenius, A. and Simons, K. (1975), *Biochim. biophys. Acta*, **415**, 26–79.

Helenius, A. and Simons, K. (1977), *Proc. natn. Acad. Sci. U.S.A.*, **74**, 529–532.

Howe, C., Blumenfeld, O.O., Lee, L.T. and Copeland, P.C. (1971), *J. Immunol.*, **106**, 1035–1042.

Howe, C. and Lee, L.T. (1969), *J. Immunol.*, **102**, 573–592.

Hunter, W.M. (1971), *Radioimmunoassay Methods* (Kirkham, K.E. and Hunter, W.K., eds.), pp. 2–23, Churchill Livingstone, Edinburgh and London.

Johansson, K.E., Blomquist, I. and Hjertén, S. (1975), *J. biol. Chem.*, **250**, 2463–2469.

Kindmark, C.-O. and Thorell, J.I. (1972), *Scand. J. clin. lab. Invest.*, **29**, Suppl. **124**, 49–53.

Kjaervig, M., Ingild, A. (1981), *Scand. J. Immunol.*, Suppl. 1, in press.

Krøll, J. (1976a), *J. Immunol. Methods*, **13**, 125–130.

Krøll, J. (1976b), *J. Immunol. Methods*, **13**, 333–339.

Lasky, R.A. and Mills, A.D. (1977), *FEBS Letters*, **82**, 314–316.

Laurell, C.B. (ed.), (1972), *Scand. J. clin. lab. Invest.*, **29**, Suppl. 124.

Laurell, C.B. (1965), *Anal. Biochem.*, **10**, 358–361.

Liljas, L., Lundahl, P. and Hjertén, S. (1974), *Biochim. biophys. Acta*, **352**, 327–337.

Løwenstein, H., Markussen, B. and Weeke, B. (1976), *Int. Arch. allergy Immunol.*, 51, 48–67.
Maire, M., Møller, J.V. and Tanford, C. (1976), *Biochemistry*, 15, 2336–2342.
Marchalomis, J.J. (1969), *Biochem. J.*, 113, 299–305.
Marchesi, V.T. (1972), In: *Methods in Enzymology* (Ginsburg, V. ed.), Vol. 28, 252–254, Academic Press, New York.
Mattsson, C., Heilbronn, E., Ramlau, J. and Bock, E. (1979), *J. Neurochem.*, 32, 301–311.
Niediech, B. (1978), *Immunochemistry*, 15, 11–12.
Nielsen, C.S. (1975), *Scand. J. Immunol.*, 4, Suppl. 2, 101–106.
Norén, O. and Sjöström, H. (1979), *J. biochem. biophys, Methods*, 7, 59–64.
Nørgaard-Pedersen, B. (1973), *Clin. chim. Acta*, 48, 345–346.
Nørgaard-Pedersen, B. and Axelsen, N.H. (1976), *Clin. chim. Acta*, 71, 343–347.
Norrild, B., Bjerrum, O.J. and Vestergaard, B.F. (1977), *Anal. Biochem.*, 81, 432–441.
O'Farrell, P.H. (1975), *J. biol. Chem.*, 250, 4007–4021.
Owen, P. and Smyth, C.J. (1976), In: *Immunochemistry of Enzymes and Their Antibodies* (Salton, M.R.J., ed.), pp. 147–202, Wiley and Sons, New York.
Owen, P. and Salton, M.R.J. (1976), *Anal. Biochem.*, 73, 20–26.
Pillion, D.J., Carter-Su, C.A., Pilch, P.F. and Czech, M.P. (1980), *J. Biol. Chem.*, in press.
Plesner, T.C. (1978), *Scand. J. Immunol.*, 8, 363–367.
Plesner, T. and Bjerrum, O.J. (1979), *Scand. J. Immunol.*, 11, 341–351.
Pødenphant, J., Bock, E., Thymann, M. and Gozes, I, (1981), *Scand. J. Immunol.*, Suppl. 11, in press.
Raftell, M. and Blomberg, F. (1974), *Eur. J. Biochem.*, 49, 31–39.
Ramlau, J. and Bjerrum, O.J. (1977), *Scand. J. Immunol.*, 6, 867–871.
Rowe, D.S. (1969), *Bull. WHO*, 40, 613–616.
Rush, R.A., Kindlev, S.H. and Udenfriend, S. (1974), *Biochem. biophys. Res. Comm.*, 61, 38–44.
Schmidt-Ullrich, R., Thompson, W.S. and Wallach, D.F.H. (1977), *Proc. natn. Acad. Sci. U.S.A.*, 74, 643–647.
Schmidt-Ullrich, R., Wallach, D.F.H. and Hendricks, J. (1975), *Biochim. biophys. Acta*, 382, 295–310.
Shivers, C.A. and James, J.M. (1967), *Immunology*, 13, 547–554.
Singer, S.J. and Nicholson, G.L. (1972), *Science*, 175, 720–731.
Slinde, E. and Flatmark, T. (1976), *Biochim. biophys. Acta*, 455, 796–805.
Steck, T.L. (1974), *J. cell Biol.*, 62, 1–19.
Svendsen, P.J. (1973), *Scand. J. Immunol.*, 2, Suppl. 1, 69–70.
Svendsen, P.J. and Axelsen, N.H. (1972), *J. Immunol. Methods*, 1, 169–176.
Takeo, K. and Kabat, E.A. (1978), *J. Immunol.*, 121, 2305–2310.
Tanford, C. and Reynolds, J. (1976), *Biochim. biophys. Acta*, 457, 133–170.
Tanner, M.J.A. and Anstee, D.J. (1976), *Biochem, J.*, 153, 265–270.
Teichberg, V.I., Sobel, A. and Changeux, J.-P. (1977), *Nature*, 267, 540–542.
Thorell, J.I. and Johansson, B.G. (1971), *Biochim. biophys. Acta*, 251, 363–369.
Uriel, J. (1971), In: *Methods in Immunology and Immunochemistry* (Williams, C.A. and Chase, M.W., ed.), Vol. 3, pp. 294–321, Academic Press, New York.

Verbruggen, R. (1975), *Clin. Chem.*, **21**, 5–43.

Vestergaard, B.F. (1975), *Scand. J. Immunol.*, **4**, Suppl. 2, 203–206.

Vestergaard, B.F. and Grauballe, P.L. (1975), *Scand. J. Immunol.*, **4**, Suppl. 2, 207–210.

Walsh, P.S., Barber, B.H. and Crumpton, M.J. (1977), *Biochem. Soc. Trans.*, **5**, 1134–1137.

Weeke, B. (1973a), *Scand. J. Immunol.*, **2**, Suppl. 1, 15–46.

Weeke, B. (1973b), *Scand. J. Immunol.*, **2**, Suppl. 1, 47–57.

Westin, M. (1976), *J. embryol. exp. Morph.*, **35**, 507–519.

Wiedmer, T. (1974), *FEBS Letters*, **47**, 260–263.

Wroblewski, H., Johansson, K.E. and Hjertén, S. (1977), *Biochim. biophys. Acta*, **465**, 275–289.

Yu, J. and Steck, T.L. (1975), *J. biol. Chem.*, **250**, 9176–9184.

6   Quantitative Methods for Studying the Mobility and Distribution of Receptors on Viable Cells

JOSEPH SCHLESSINGER and ELLIOT L. ELSON

Membrane Receptors: Methods for Purification and Characterization
(*Receptors and Recognition*, Series B, Volume 11)
Edited by S. Jacobs and P. Cuatrecasas
Published in 1981 by Chapman and Hall, 11 New Fetter Lane, London EC4P 4EE
© Chapman and Hall

6.1 POSSIBLE SIGNIFICANCE OF RECEPTOR MOBILITY

The rapid two-dimensional motion of membrane receptors provides an efficient mechanism for communication among various receptors and for assembly of multi-molecular structures in the plane of the membrane. There is now good evidence that the lateral motion of receptor molecules plays an important physiological role in a number of systems. A relationship between receptor motion and aggregation and biological response has been most convincingly demonstrated in the degranulation of mast cells and basophils. (Ishizaka and Ishizaka, 1971). It was shown that the F_c receptor for immunoglobulin E (I_gE) is monovalent and that the IgE-receptor complex can diffuse in the plane of the membrane with diffusion coefficient $D \sim 3 \times 10^{-10} \, cm^2 \, sec^{-1}$ (Mendoza and Metzger, 1976; Schlessinger et al., 1976b). The degranulation of mast cells and basophils can be provoked in three different ways; by cross linking IgE-receptor complexes with anti-IgE antibodies (Sirganian et al., 1975); by binding chemically cross-linked IgE dimers to receptors (Segal et al., 1974); by crosslinking the (unoccupied) F_c receptors with divalent anti-receptor antibodies (Isersky et al., 1978). These results indicate that aggregation of the IgE receptors (presumably due to diffusional encounters between receptors) provides the signal for triggering degranulation, and that the 'unit signal' results from forming a dimer of receptors.

Receptor motion seems also to play an important role in the activation of hormonal responses (Perkins, 1973; Cuatrecasas, 1974; De Meyts et al., 1976). The fact that different hormones which bind to various different cell surface receptors can activate the same pool of adenylate cyclase suggests that these receptors are mobile (Cuatrecasas, 1974). Furthermore, β-adrenergic receptors from turkey erythrocytes can activate the adenylate cyclase of other cell types when the receptor-bearing and the enzyme-bearing cells are fused together with Sendai virus (Orly and Schramm, 1976). Tolkovsky and Levitzki have proposed that the lateral motion of β-adrenergic receptors plays a role in the activation of adenylate cyclase. They have shown that the coupling of turkey erythrocyte cyclase with the receptor occurs via a 'collisional coupling' mechanism whereas the adenosine receptor is permanently coupled to the enzyme in the same cell (Tolkovsky and Levitzki, 1978a and b). Furthermore, modulation of the fluidity of the plasma membrane of turkey erythrocytes by insertion of cis-vaccinic acid can affect the activation of adenylate cyclase (Simon et al., 1978). The rate of the activation of the enzyme by L-adrenaline depends on membrane fluidity while the adenosine-induced rate of enzyme activation is independent of membrane fluidity. Hirata et al., observed that the β-adrenergic agonist L-isoproterenol increases phospholipid methylation which causes a decrease in membrane viscosity (1979). This increased the lateral motion of

159

the receptor molecule which, in turn, enhanced the coupling between the receptor and the adenylate cyclase. It seems that activation of the cyclase by adrenalin is sensitive to changes in the lipid composition (Simon *et al.*, 1978; Hirata *et al.*, 1979). This is probably the consequence of lipid effects on the micro-environment around the receptor and the enzyme and the degree of coupling between them. The hormone-induced phospholipid methylation could function as a regulatory step by which hormone binding modulates membrane viscosity which affects the coupling between various membrane receptors.

In the last year several laboratories have shown that lateral motion and clustering of the complexes of insulin and epidermal growth factor with their receptors play a role in their mechanism of action. It is believed that these two hormones act directly on the plasma membrane of their target cells, though internalization has been demonstrated (Terris and Steiner, 1975; Steiner, D.F., 1977; Goldfine *et al.*, 1977; Carpenter and Cohen, 1978) and postulated to yield a species or fragment (Das and Fox, 1978) which is active inside the cell. Many substances have been proposed as second messengers of insulin (Czech, 1977). At present there is insufficient evidence for any of the proposed second messengers as mediators of the diverse responses of the hormone.

The lateral motion and distribution of fluorescent analogues of insulin and EGF on cultured fibroblasts has recently been studied. It was shown that insulin and EGF are initially mobile and homogeneously distributed over the cell surface. Within a few minutes, at 23°C or 37°C, the hormone receptor complexes aggregate into visible immobile patches on the cell surface and shortly thereafter the patches became internalized into endocytic vesicles. The hormone-induced aggregation can be divided into two processes: a local aggregation which involves only a few receptor molecules, and a global aggregation which involves motion over distances of several microns to form visible patches which contain hundreds of receptor molecules. It has been suggested that the local aggregation process is related to the rapid membrane response to insulin and EGF and initiates the process of thymidine incorporation activated by EGF (Schlessinger, 1979; Shechter *et al.*, 1979). The global aggregation is related to receptor down regulation and precedes the internalization and degradation of insulin and EGF (Schlessinger, 1979). Several other examples suggest the involvement of receptor lateral motion in biologically important processes. Lymphocytes can be stimulated to differentiate and proliferate by bivalent anti-immunoglobulin and lectin which induce the aggregation of their mobile receptors (Cunningham *et al.*, 1976; Siekman *et al.*, 1978). Lateral motion of cell-surface molecules is associated with the processes of endocytosis (Tsan and Berlin, 1971) and exocytosis (Chi *et al.*, 1976), and the release of vesicles at the neuromuscular junction (Hevser, 1979).

Lateral motion of plasma-membrane components may also be involved in the assembly of higher-order surface structures. Possible examples include the formation of patches of acetylcholine receptors (Axelrod *et al.*, 1976b) and sites of attachment of internal fibrous components (cytoskeleton) to the cell membrane (Geiger and

Singer, 1979). Direct evidence for the role of mobility in these processes is still lacking, however. An interesting model is provided by the assembly of buds of internal viruses (Garoff and Simons, 1974; Reidler *et al.*, 1979).

These various examples demonstrate the need for direct methods to study the lateral mobility of specific cell surface components and to detect their precise localization on viable cells.

This chapter describes the application of two new simple methods for monitoring the distribution and mobility of very low levels of fluorescently labeled markers on viable cells. Localization of fluorescent species on the cells is accomplished with a newly developed, highly sensitive video intensification microscopy system which permits the visualization of light at exceedengly low levels. The second method, denoted fluorescence photobleaching recovery, gives quantitative information about lateral mobility of receptors on the cell surface.

6.2 QUANTITATIVE DETERMINATION OF THE LATERAL MOTION OF MEMBRANE COMPONENTS

Recently several groups have begun to measure the lateral diffusion of membrane components with fluorescence photobleaching recovery (FPR) methods. In all versions of these methods the lateral diffusion is measured by inducing an inhomogeneous distribution of fluorescent molecules on the cell membrane. A small region of the cell membrane which is labeled with mobile fluorescent molecules is illuminated with a focused laser beam (radius $\sim 1 \mu m$). A brief and intense pulse of light causes an irreversible photobleaching of some of the fluorophores in the illuminated region. The diffusion coefficients are determined by measuring the rate of recovery of fresh fluorophores into the bleached area (which results from the lateral motion of the labeled molecules in the plane of the membrane). The fluorescence from the bleached area is excited by an attenuated beam in order to avoid photobleaching during the course of fluorescence recovery. In our version of this method we employ a Gaussion focussed laser beam both for bleaching, and after attenuation (of 10^3 or 10^4) for monitoring diffusion of fluorescent molecules into the bleached area.

We assume that the light-induced conversion of the fluorophores to non-fluorescent species can be described as an irreversible first order reaction with a rate constant $\alpha I(r)$ (Axelrod *et al.*, 1976). The concentration of the unbleached fluorophores $C(r, t)$ at position r and time t can be calculated from

$$\frac{dC(r, t)}{dt} = -\alpha \cdot I(r) \cdot C(r, t) \tag{1}$$

where $I(r)$ is the bleaching intensity and α is the bleaching rate constant. For a bleaching pulse which lasts a time interval T (short compared to the recovery time) the fluorescence concentration profile at the beginning of the recovery phase $(t = 0)$ is given by:

$$C(r, 0) = C_0 \cdot \exp \cdot (-\alpha T \cdot I(r)) \tag{2}$$

where C_0 is the initial uniform fluorophore concentration. The 'amount' of bleaching induced in time T is expressed by a parameter K:

$$K = \alpha \cdot T \cdot I(0) \tag{3}$$

where $I(0)$ is the (central, maximum) bleaching light intensity. For a Gaussian profile $I(r)$ is given by

$$I(r) = (2P_0 \, \pi^{-1} \, W^{-2}) \exp(-2r^2 \, W^{-2})$$

$$I(0) = 2P_0 \, \pi^{-1} \, W^{-2} \tag{4}$$

where W is the half width at e^{-2} height and P_0 is the total laser power. Equation (1) was solved for various K values (Axelrod *et al.*, 1976). The recovery by diffusion of fluorescence, $F_K(t)$, from photobleaching for a Gaussian profile is given by

$$F_K(t) = (q \, P_0 \, C_0/A) \sum_{n=0}^{\infty} \frac{(-K)^n}{n!} \cdot \frac{1}{1 + n(1 + 2t \, \tau_D^{-1})} \tag{5}$$

where q is the product of the quantum efficiencies for light absorption, detection and emission, and A is the attenuation factor equal to the bleaching intensity. τ_D is the diffusion time constant.

The fluorescence intensity just after bleaching is a function of K alone and is given by

$$F_K(0) = \frac{q \, P_0 \, C_0}{A} \cdot \frac{(1 - e^{-K})}{K} \tag{6}$$

when $K \ll 1$ the fluorescence recovery curves can be represented in the simple form

$$F(t) = F(0) \cdot (1 + t \, \tau_D^{-1})^{-1} \tag{7}$$

The diffusion time constant τ_D is equal to $W^2 \, (4D)^{-1}$. In many experiments we have calculated diffusion coefficient from the time required for half of the observed fluorescence recovery to occur $(\tau_{1/2})$: $\tau_{1/2}$ equals $\tau_D \, \gamma_D$. Axelrod *et al.*, have calculated the values of γ_D for various values of K. Theory was also developed for uniform flow or for simultaneous diffusion and flow (Axelrod *et al.*, 1976a). If some of the fluorophore in the illuminated region is immobile, then the asymptote of the fluorescence recovery after bleaching $F_K(\infty)$ will be less then F_K (initial). The fraction of the mobile fluorophores ($\% R$) is given by

$$\% R = (F_K(\infty) - F_K(0))/(F_K(\text{initial}) - F_K(0)) \tag{8}$$

The diffusion coefficient D is calculated from $\tau_{1/2}$, W and γ_D. γ_D is calculated for each K value. The mobile fraction $\% R$ is calculated according to equation (8). Recently Smith and McConnell (1978) employed another version of fluorescence photobleaching recovery in order to measure the diffusion coefficients of fluorescent lipid embedded in artificial lipid bilayers. Instead of photobleaching a small circulary symmetric region on the membrane, they have photobleached a periodic pattern of parallel strips. The periodic pattern is photographed at various times after photobleaching and a densitometer trace is recorded for each photograph. Smith and McConnel have shown that the amplitude of the periodic pattern decays exponentially and that

$$D = 1/a^2 \tau \qquad (9)$$

where a is the spatial frequency of the pattern and is equal to $2\pi/P$, where P is the period of the pattern and τ is the life time of the exponential decay. Currently this version of fluorescence recovery after photobleaching has been primarily employed only for studying the motion of fluorescent lipid analogues embedded in artificial lipid bilayers (Smith and McConnell, 1978).

We and others have used the FPR method to study the motion of various membrane molecules of many cell types under a variety of physiological conditions. In these studies it has been demonstrated that FPR has many favorable features for measuring the mobilities of various surface components on living cells.

Because the observation region is small (~ 1 μm radius), it is possible to measure the mobility of components in different regions on the same cell membrane and to compare their values. This capability has been important in some studies: Axelrod *et al.* (1976) have shown that acetylcholine receptors on cultured myotubes appear in two classes: a mobile, diffuse population and an immobile population which is specifically localised in patches on the plasma membrane. In another study we have shown that local binding of platelets, coupled to the lectin conconavalin A (con A), onto one part of the cell surface can retard the diffusion of surface antigens on other parts of the plasma membrane of the same cell (Schlessinger *et al.*, 1977c). Thus, cross linking and immobilization of Con A receptors on one part of the cell can induce a global decrease in receptor motion (Schlessinger *et al.*, 1977c). In collaboration with S. de Laat we have shown that lateral mobilities of membrane lipids and antigens is increased specifically in the outgrowing neurites of neuroblastoma cells. This indicates the existence of a topographical heterogeneity in the cell membrane of differentiating neuroblastoma cells, and suggests that the more fluid lipid domains in the cell membrane are located in the neurites (de Laat *et al.*, 1979).

In FPR a fluorescence microscope enables a fairly precise localization of the observation region relative to visible cellular features. The narrow depth of field of the microscope allows a determination that the fluorescent marker is on or near the cell surface. Furthermore, by using two different fluorescent markers (like rhodamine and fluorescein) it is possible to measure the effect of one membrane molecule (or matrix) on the mobility of a second nearby molecule. We have used this special feature of FPR to demonstrate that the presence of immobile fibronectin (CSP/LETS) fibers does not impede the diffusion of a lipid probe, a ganglioside analogue and various surface antigens on chicken embryo fibroblasts. Therefore the fibronectin fibrils do not seem to form a 'barrier' across the lipid matrix of the plasma membrane. In contrast, Con A, which binds to fibronectin, is highly immobile in areas which are rich in this protein and highly mobile in areas which are poor in fibronectin (Schlessinger *et al.*, 1977b).

Another common problem in studies on membrane dynamics originates from the experimentally observed fact that various fluorescent lipid probes become rapidly internalised. Therefore their motion does not reflect dynamic aspects of the plasma membrane. Employing FPR we have been able to show that 3, 3' dioctadecylindo-carbocyanine iodide (diI), and a fluorescent analogue of the ganglioside GM_1, are both good markers of the lipid phase of the plasma membrane. This is in contrast to many other lipophylic dyes which become rapidly internalised (Schlessinger *et al.*, 1977a; Schlessinger and Elson, 1980).

A major advantage of the FPR method is its ability to yield diffusion rates in individual cells under physiological conditions. This permits a correlation of changes in cellular physiological state and morphology with dynamic processes in the plasma membrane. Examples include activation of histamine release from mast cells and basophils (Ishizaka and Ishizaka, 1971; Mendoza and Metzger, 1976; Schlessinger *et al.*, 1976b; Sirganian *et al.*, 1975; Segal *et al.*, 1974; Isersky *et al.*, 1978), the response of various cell types to insulin and epidermal growth factor (Schlessinger *et al.*, 1978a and b; Schlessinger, 1979; Shechter, 1979), and the assembly of specialized surface structures such as patches of acetylcholine receptors on myotubes (Axelrod *et al.*, 1976b) and viral buds on cells infected with enveloped viruses (Garoff and Simons, 1974; Reidler *et al.*, 1979).

Edidin and Johnson have studied the motion of diI and surface antigens during the process of egg fertilization. They have shown that both lipid molecules and surface antigens become less mobile upon fertilization of mouse eggs (Johnson and Edidin, 1978). This reduction in membrane fluidity and in receptor motion could be related to the prevention of polyspermy (Johnson and Edidin, 1978).

In collaboration with S. de Laat we have measured the diffusion coefficients of lipid probes and surface antigens throughout the cell cycle of neuroblastoma cells. We have shown that lipid diffusion reaches a minimum in mitosis, increases 2- to 3-fold during G1, remains constant at maximum value during S and decreases again shortly before mitosis. However, the diffusion of surface antigens under the same conditions is different; they also show minimum value for diffusion in mitosis followed by a similar rise in G1. But through S and G_2 the diffusion of membrane antigens gradually decreases. These results (de Laat *et al.*, 1980) indicate that lipid diffusion is controlled by membrane viscosity, while antigen diffusion is effected by additional interactions in the plasma membrane.

Recently Maeda *et al.* (1979) have demonstrated that treatment with sendai viruses effected the lateral motion of cell surface components on cultured human KB cells and mouse 3T3 cells. They have shown that interaction with sendai virus increases the diffusion coefficients of Con A binding sites and β_2-microglobulin by 2– to 3-fold. In contrast, visus treatment slightly reduced the diffusion coefficient of a lipid analogue (Maeda *et al.*, 1979). These results indicate that the effects on the motion of membrane proteins are not due to changes in membrane fluidity but are the consequence of viral effects on other membrane interactions.

One of the main features of FPR is its versatility. A wide range of mobilities is accessible. Diffusion coefficients can be measured over the range from 10^{-12} to 10^{-6} cm^2 s^{-1}. The specificity of the diffusion measurement is determined by the fluorescently labeled marker which is used. Up to now these have included a lipid probe (de Laat *et al.*, 1979; Schlessinger *et al.*, 1977a; Johnson and Edidin, 1978; de Laat *et al.*, 1980; Schlessinger *et al.*, 1976a; Elson *et al.*, 1976; Elson and Schlessinger, 1979), a ganglioside analogue (Schlessinger and Elson, 1979), unselected surface proteins (Schlessinger *et al.*, 1977a; Edidin *et al.*, 1976) and antigens (Schlessinger *et al.*, 1977c; Johnson and Edidin, 1978; de Laat *et al.*,1980), Con A binding sites (Schlessinger *et al.*, 1976a; Zagyansky and Edidin, 1976; Jacobson *et al.*, 1978) acetylcholine receptors (Axelrod *et al.*, 1976b) IgE receptors (Schlessinger *et al.*, 1976) and insulin and EGF receptors (Schlessinger *et al.*, 1978a). In all these studies the lateral diffusion over micron distances is measured directly. It is not necessary to infer rates of macroscopic diffusion from measurements of microscopic motions as in magnetic resonance or fluorescence depolarization.

6.3 LOCALIZATION OF MOLECULES ON CELLS BY IMAGE-INTENSIFICATION TECHNIQUES

Fluorescence microscopy has been used extensively for studying the localization of fluorescently labeled molecules in cells. These studies have yielded important information on the topographical distribution of various molecules on the cell surface, inside the cytoplasm, and in the nucleus. The specificity of the labeling was usually achieved by immunological methods. Antibodies against cellular antigens were prepared, then the cells were treated with these antibodies and their distribution was determined by a second fluorescent-labeled antibody directed against the primary one. In this way indirect immunofluorescence microscopy was used to study the localization of various cellular antigens. Similarly, other cellular components were localised after being labeled with fluorescently tagged lectins, toxins, hormones, etc.

One of the major shortcomings of conventional fluorescence microscopy is that the fluorescent dye which is attached to the proteins can be visualised for relatively short periods of time. This is because the external illumination required to observe them by eye causes rapid fading of the fluorescence intensity. For some fluorophores the light-induced fading is so rapid that it is impossible to record their image, even on a very sensitive film.

One way to overcome the problem of photobleaching is to decrease the intensity of the excitation light (this will reduce the rate of the photobleaching process) and then use a sensitive mean for monitoring low levels of light. Under these conditions the total number of the emitted quanta provided by the specimen becomes small, and methods for 'image intensification' become necessary.

All image intensifiers have the following common features: the photons which are emitted from the specimen are converted to electrons. The number of the electrons is multiplied and their energy is increased. The electrons are kept in spatial focus by means of electric and magnetic fields and eventually their energy is converted to a visual image which can be photographed or video monitored.

The main advantages of image intensification over conventional microscopy is summarised as follows:

(1) Very low excitation intensity permits continuous recording of fluorescent markers on viable cells for more than 1 h without photobleaching.

(2) The high sensitivity of the camera allows detection of a few hundred fluorescent molecules per μm^2.

(3) The continuous observation of the cells under physiological conditions allows observation of specific markers over periods as long as a few days.

(4) The low excitation intensity permits recording of cells without causing light-induced cellular damage.

The technology of image intensification has been developed over the past 20 years and is still in a state of continuing development. It has been applied in various fields of science but very rarely in the field of biology. Reynolds first introduced image intensifiers for studying biological systems. He used these devices to observe bioluminescence and fluorescence (Reynolds, 1972 and 1968; Reynolds and Botos, 1970). Up to now image intensification has primarily been employed in studies of cells emitting low levels of fluorescence or bioluminescence. Another application of image intensification involves the collection and processing of images of living cells which are extremely photosensitive and therefore need to be studied under low light levels (Dvorak *et al.*, 1976).

The first application of image intensification in biology was reported in 1967. Eckert and Reynolds (1967) visualised the bioluminescence emitted from *Noctiluca miliaris* after electrical or mechanical stimulation. They have shown that the flashes emitted from this single-cell dinoflagellate arise from a large number of microscopic sources of sizes ranging from 0.5 to 1.5 μm located near the cell periphery. The use of an image intensifier permitted photographic recording of these microflashes. More examples from the field of bioluminescence are reviewed by Reynolds (1972).

Rose and Loewenstein have studied the distribution of Ca^{2+} in the cytoplasm of cells of isolated chironomus salivary glands. The cells were microinjected with the Ca^{2+}-sensitive luminescent protein aequorin. The aequorin light emission, which is proportional to Ca^{2+} concentration, was viewed and recorded through a microscope with the aid of an image intensifier coupled to a TV camera (Rose and Loewenstein, 1975). They found that the diffusion of Ca^{2+} through the cytosol is so constrained that a rise in the cytoplasmic concentration produced by local injection is confined to the vicinity of the site of injection. Employing a similar approach, Taylor *et al.*, have studied the localization of Ca^{2+} after the microinjection of aequorin into giant ameoba (1975). They found that luminescence is induced in parallel with cellular movements when sufficiently strong electrical or mechanical stimuli are applied. Also in ameoba, as in salivary gland cells, Ca^{2+} seems to be highly constrained by a sequestering system.

Sedlacek *et al.* (1976) were the first to apply image intensification in order to visualise fluorescent-labeled markers on the cell surface. Their apparatus was composed of a conventional fluorescence microscope, a television camera equipped with an image intensifier, a television monitor and a video recorder. They report that with this equipment it is possible to reduce the power of the excitation light and therefore follow the fluorescence of cell membrane immunoglobulins or lymphocytes for more than 30 minutes.

Image intensification seems to be especially useful in the field of endocrinology. Hormone receptors usually appear in very low numbers on the cell surface. Therefore it is difficult to photograph the distribution of fluorescently labeled hormones on the cell membrane with conventional films. In order to overcome this problem, Shechter *et al.*, have prepared highly fluorescent analogues of insulin and epidermal growth factor (7–8 rhodamines per hormone molecule). These analogues compete with native hormones for the receptor and retain bioactivity (Shechter *et al.*, 1978). We have added the fluorescent hormones to cells and visualised their distribution on the cells with a SIT (silicon intensified target) camera and more recently with an ISIT (intensified silicon intensified target) camera. These video cameras are 10^4 to 10^5 times more sensitive then conventional TV cameras or conventional photographic film. Initially the fluorescent hormones were homogeneously distributed over the cell surface (Schlessinger *et al.*, 1978a and b; Schlessinger, 1979). Within a few minutes at 23°C or 37°C the hormone receptor complexes aggregated into patches on the cell surface and within 30 min at 37°C much of the aggregated hormone became internalized into endocytic vesicles. The aggregation process did not require metabolic energy but internalization of hormones requires metabolic energy. Using fluorescein labeled α_2-macroglobulin and rhodomine-labeled insulin or EGF (Maxfield *et al.*, 1978), all three proteins were seen to collect at the same patches (coated regions) on the cell surface and were internalised within the same vesicles (coated vesicles). Recently we have reported that a cyanogen bromide-cleaved EGF derivative binds to its membrane receptors on 3T3 cells but cannot elicit thymidine incorporation. This analogue appeared to be homogeneously distributed over the cell membrane. However, when crossed-linked with anti-EGF antibodies, bioactivity was restored and visual patches appeared (Shechter *et al.*, 1979).

Taylor and Wang have prepared skeletal muscle actin labeled with 5-iodoacetamidofluorescein. The fluorescent actin has been microinjected into living sea urchin eggs during early development (Taylor and Wang, 1979). The distribution of fluorescence following fertilization and during the first cell division was monitored with a SIT camera. In unfertilised eggs the fluorescence of actin was uniformly distributed inside the egg. Following fertilization of eggs containing labeled actin, regions close to the plasma membrane showed an increase of fluorescence intensity which was concurrent with the elevation of the fertilization membrane. The appearance of fluorescence streaks during cytokinesis raised the possibility that longitudinally oriented microfilament bundles were involved in the process of cell cleavage.

Dvorak *et al.* (1976) have recently reported on a low-light-level video equipment for monitoring the fluorescence of reduced nicotinamide adenine dinucleotide (NADH) in the cerebral cortex of the exposed brain. The goal of the study is to relate oxidative metabolism in the brain as indicated by NADH fluorescence. Initial results using this instrument indeed indicate that NADH Fluorescence, O_2 consumption and potassium kinetics are closely interrelated. The low-light-level video system seems to be a useful method for studing redox changes during brain activity *in vivo*.

The same group is using this equipment for studing the interaction with the red blood cell of the parasite which causes malaria. The use of image intensification is particularly important in this system because the cells are extremely sensitive to light. The results of the study show that the invasion by malaria merozoites consists of attachment of the anterior end of the parasite to the erythrocyte, deformation of the erythrocyte and entry of the parasite by erythrocyte membrane invagination (Dvorak *et al.*, 1975).

Although image intensification has so far had a limited use in cell biology, it seems likely that in the future it will be an important addition to conventional techniques.

REFERENCES

Axelrod, D., Koppel, D.E., Schlessinger, J., Elson, E.L. and Webb, W.W. (1976a), *Biophys. J.*, **16**, 1055–1069.

Axelrod, D., Ravdin, P., Koppel, D.E., Schlessinger, J., Webb, W.W., Elson, E.L. and Podleski, T.R. (1976b), *Proc. natn. Acad. Sci. U.S.A.*, **73**, 4594–4598.

Carpenter, G. and Cohen, S. (1976), *J. cell Biol.*, **71**, 159–171.

Chi, E.Y., Lagunoff, D. and Koehler, J.K. (1976), *Proc. natn. Acad Sci. U.S.A.*, **73**, 2823–2827.

Cuatrecasas, P. (1974), *Ann. Rev. Biochem.*, **43**, 169–214.

Cunningham, B.A., Sela, B.A., Yahara, I. and Edelman, G.M. (1976), In: *Mitogens in Immunobiology* (Oppenheim, J.J. and Rosentreich, D.L., eds.), pp. 13–30, Academic Press, New York.

Czech, M.P. (1977), *Ann. Rev. Biochem.*, **46**, 359–384.

Das, M. and Fox, C.F. (1978), *Proc. natn. Acad Sci. U.S.A.*, **75**, 2644–2648.

DeMeyts, P., Bianco, A.R. and Roth, J. (1976), *J. biol. Chem.*, **251**, 1877–1888.

Dvorak, J.A., Miller, L.H., Whitehouse, C.W. and Shiroishi, T. (1975), *Science*, **183**, 748–750.

Dvorak, J.A., Miller, L.H., Whitehouse, C.W. and Shiroishi, T. (1975), **183**, 748–750.

Dvorak, J.A., Schuette, W.H. and Whitehouse, W.C. (1976), *Spic, Low light level devices*, **78**, 155–160.

Eckert, R. and Reynolds, G.T. (1967), *J. gen. Physiol.*, **50**, 1429–1458.

Edidin, M., Zagyansky, Y. and Larder, T.J. (1976), *Science*, **196**, 466–468.

Elson, E.L. and Schlessinger, J. (1979), In: *The Neurosciences: Fourth Study Program.* (Schmitt, F.O. and Worden, F.G., eds.), pp. 691–701, MIT Press, Cambridge, Mass.

Elson, E.L., Schlessinger, J., Koppel, D.E., Axelrod, D. and Webb, W.W. (1976), In: *Membranes and Neoplasia: New Approaches and Strategies*, (Marchesi, U.T., ed.), pp. 137–147, New York, Allan R. Liss.

Garoff, H. and Simons, K. (1974), *Proc. natn. Acad. Sci. U.S.A.*, **71**, 3988–3992.

Geiger, B. and Singer, S.J. (1979), *Cell*, **16**, 213–222.

Goldfine, I.D., Smith, G.J., Wong, K.Y. and Jones, A.L. (1977), *Proc. natn. Acad. Sci. U.S.A.*, **74**, 1368–1372.

Heuser, J.F. (1979), *The Neurosciences: Fourth Study Program*, (Schmitt, F.O. and Worden, F.G., eds.), MIT Press, Cambridge, Mass.

Hirata, F., Strittmatter, W.J. and Axelrod, J. (1979), *Proc. natn. Acad. Sci. U.S.A.*, **76**, 368–372.

Hirata, F., Strittmatter, W.J. and Axelrod, J. (1979), **76**, 368–372.

Isersky, H., Taurog, J., Poy, G. and Metzger, H. (1978), *J. Immunol.*, **12**, 549–558.

Ishizaka, K. and Ishizaka, T. (1971), *Ann. N.Y. Acad. Sci.*, **190**, 443–456.

Jacobson, K., Wu, G. and Poste, G. (1976), *Biochim. biophys. Acta.*, **433**, 215–222.

Johnson, M. and Edidin, M. (1978), *Nature*, **272**, 448–451.

Koppel, D.E., Axelrod, D., Schlessinger, J., Elson, E.L. and Webb, W.W. (1976), *Biophys. J.*, 1315–1329.

de Laat, S.W., Van der Saag, P.T., Elson, E.L. and Schlessinger, J. (1980a), *Proc. natn. Acad. Sci. U.S.A.*, **77**, 1526–1528.

de Laat, D.W., Van der Saag, P.T., Elson, E.L. and Schlessinger, J. (1979), *Biophys. biochim. Acta*, **558**, 247–250.

Maeda, T., Eldrige, C., Toyama, S., Ohnishi, S.-I., E.L. and Webb, W.W. (1979), *Exp. cell Res.*, **123**, 333–343.

Maxfield, F., Schlessinger, J., Shechter, Y., Pastan, I. and Willingham, M.C. (1978), *Cell*, **14**, 805–810.

Mendoza, G. and Metzger, H. (1976), *Nature*, **264**, 548–549.

Orly, J. and Schramm, M. (1976), *Proc. natn. Acad. Sci., U.S.A.*, **73**, 4410–4416.

Perkins, J.P. (1973), *Adv. cyc. nucleotide Res.*, **34**, 1–64.

Reidler, J., Lenard, J., Keller, P., Schlesinger, M., Johnson, D. and Elson, E.L., in preparation. Also in J. Reidler (1979), Ph. D. Thesis, Cornell University.

Reynolds, G.T. (1968), *Adv. opt. electron. Microsc.*, **2**, 1–40.

Reynolds, G.T. (1972), *Quarterly Rev. Biophys.*, **5**, 295–347.

Reynolds, G.T. and Botos, P. (1970), *Biol. Bull. mar. biol. Lab. Woods Hole.*, **139**, (2), 432.

Rose, B. and Loewenstein, W.R. (1975), *Science*, **190**, 1204–1206.

Schlessinger, J. (1979), In: *Physical Chemical Aspects of Cell Surface Events in Cellular Regulation* (Delisi, C. and Blumenthal, R., eds.), pp. 89–111, Elsevier Press, New York.

Schlessinger, J., Axelrod, D. Koppel, E.E., Webb, W.W. and Elson, E.L. (1977a), *Science,* **195**, 307–309.

Schlessinger, J., Barak, L.S., Hammes, G.G., Yamada, K., Pastan, I., Webb, W.W. and Elson, E.L. (1977b), *Proc. natn. Acad. Sci., U.S.A.,* **74**, 2909–2913.

Schlessinger, J. and Elson, E.L. (1979), In: *Biophysical Methods,* a volume in *Methods of Experimental Physics,* (Ehreinstein, G. and Lecar, H., eds.), Academic Press, New York, in press.

Schlessinger, J., Elson, E.L., Webb, W.W., Yahara, I. Rutishauser, U. and Edelman, G.M. (1977c), *Proc. natn. Acad. Sci., U.S.A.,* **74**, 1110–1114.

Schlessinger, J., Koppel, D.E., Axelrod, D., Jacobson, K., Webb, W.W. and Elson, E.L. (1976a), *Proc. natn. Acad. Sci., U.S.A.,* **73**, 2409–2413.

Schlessinger, J., Shechter, Y., Cuatrecasas, P. Willinghem, M.C. and Pastan, I. (1978a), *Proc. natn. Acad. Sci., U.S.A.,* **75**, 5353–5357.

Schlessinger, J., Shechter, Y., Willingham, M.C. and Pastan, I. (1978b), *Proc. natn. Acad. Sci., U.S.A.,* **75**, 2659–2663.

Schlessinger, J., Webb, W.W., Elson, E.L. and Metzger, H. (1976b), *Nature,* **264**, 550–551.

Sedlacek, H.H., Gundlach, H. and Ax, W. (1976), *Behring Inst. Mitt.,* **59**, 64–70.

Segal, D., Taurog, J. and Metzger, H. (1977), *Proc. natn. Acad. Sci., U.S.A.,* **74**, 2293–2297.

Shechter, Y., Hernaez, L., Schlessinger, J. and Cuatrecasas, P. (1979), *Nature,* **278**, 835–838.

Shechter, Y., Schlessinger, J., Jacobs, S., Chang, K.J. and Cuatrecasas, P. (1978), *Proc. natn. Acad. Sci., U.S.A.,* **75**, 2135–2139.

Sieckmann, E.E., Asofsky, R., Mosier, D.E., Zitron, I.M. and Paul, W.E. (1978), *J. exp. Med.,* **147**, 814–829.

Simon, G., Hanski, E., Braun, S. and Levitzki, A. (1978), *Nature,* **274**, 394–397.

Sirganian, R.P., Hook, W.A. and Levine, B.R. (1975), *Immunochemistry,* **12**, 149–157.

Smith, B.A. and McConnell, A.M. (1978), *Proc. natn. Acad. Sci., U.S.A.,* 2759–2763.

Steiner, D.F. (1977), *Diabetes,* **26**, 322–340.

Taylor, D.L., Reynolds, G.T. and Allen, R.D. (1975), *Biol. Bull.,* **149**, 448.

Taylor, D.L. and Wang, Y.L. (1979), *Proc. cell Biol. Meeting, Galvaston, Texas.*

Terris, S. and Steiner, D.F. (1975), *J. biol. Chem.,* **250**, 8389–8398.

Tolkovsky, A.M. and Levitzki, A. (1978a), *Biochemistry,* **17**, 3795–3810.

Tolkovsky, A.M. and Levitzki, A. (1978b), *Biochemistry,* **17**, 3811–3817.

Tsan, M.-G. and Berlin, R.D. (1971), *J. exp. Med.,* **134**, 1016–1035.

Zagyansky, Y. and Edidin, M. (1976), *Biochim. biophys. Acta,* **433**, 209–214.

7 Somatic Genetic Analysis of Hormone Action

GARY L. JOHNSON, PHILIP COFFINO
and HENRY R. BOURNE

Acknowledgements

The work described here was supported in part by grants from the National Institutes of Health (GM 16496, HL 06285, GM 00001, CA 23218) and the National Science Foundation (PCM 77–14397, PCM 78–07382). P.C. is the recipient of a Research Career Development Award from the National Institutes of Health (GM 00308). G.L.J. is the recipient of a Research Fellowship (GM 06849) from the National Institutes of Health.

Membrane Receptors: *Methods for Purification and Characterization*
(*Receptors and Recognition,* Series B, Volume 11)
Edited by S. Jacobs and P. Cuatrecasas
Published in 1981 by Chapman and Hall, 11 New Fetter Lane, London EC4P 4EE
© Chapman and Hall

Most of our current understanding of hormone action derives from experiments that directly extend the methods of classical biochemistry (see Chapters 1–5). More recently, immunologic techniques have provided new ways of detecting and separating receptors and other molecules that mediate hormone action (see Chapter 6). In this chapter we review a third approach, based on concepts and techniques of somatic genetics, that complements the other approaches in elucidating the molecular machinery by which cells detect and respond to hormonal signals.

Complete understanding of this molecular machinery requires that its component parts be separated, characterized, and then recombined in functional form. Genes specify these components and are mutable. By taking advantage of the precision of genetic processes and their occasional exploitable imprecisions, the genetic approach offers a powerful alternative to investigative techniques that depend on physio-chemical or antigenic differences between proteins.

This review will not undertake to explore completely the implications of genetics for understanding hormone action, including the knowledge gained from heritable disorders affecting the generation, structure, and action of hormones in man and other animals. Instead, the scope of this review will be confined to the use of somatic genetics for understanding hormone action. Indeed, we will focus primarily on experiments performed with a single cell type, the mouse S49 lymphoma. Aside from the fact that this is the subject we know most about, the focus on a single cell is justified by the existence of a considerable corpus of information. Such information is obtainable partly because the S49 cell line has characteristics which make it an almost ideal genetic system for investigation of the actions of gluco-corticoid hormones and agents that act by stimulating the synthesis of cyclic adenosine 3′, 5′-monophosphate (cAMP). The most important of these characteristics is that cAMP and glucocorticoids kill S49 cells.

The late Gordon M. Tomkins developed the notion that the cytocidal effects of glucocorticoids and cAMP on S49 cells could be exploited to select mutant cells that could be used to illuminate the actions of hormones on a molecular level. In doing so, Tomkins frankly sought to imitate earlier exploitation of microbial mutants to explain the functions of individual gene products and their interactions with one another in mediating complex cellular processes. As a result, the S49 cell provides the most fully developed example of genetic analysis of hormone action in mammalian cells. The genetic approach is now being applied in a variety of other mammalian cells, including neuroblastoma (Simantov and Sachs, 1975), a macrophage tumor line (Rosen et al., 1979), the Y-1 adrenal carcinoma (Gutman et al., 1978), and Chinese hamster ovary (CHO) cells (Pastan and Willingham, 1978).

7.1 A STRATEGY FOR DRAWING INFERENCES FROM GENETIC EXPERIMENTS

Gene mutation can allow certain kinds of inferences regarding the roles of peptide gene products in cell regulation. In principle, these inferences can also be drawn from results of exhaustive biochemical investigation, exemplified by the elucidation of the enzymes involved in metabolic pathways such as glycolysis or the tricarboxylic acid cycle. Genetic analysis may provide significant advantages, however, particularly when it is difficult or impossible, using biochemical procedures, to unravel a complex process that is best studied in intact cells. In such cases, genetics has proved most useful in allowing two kinds of inferences:

(1) The normal functions of a peptide — e.g., a membrane receptor protein — can be inferred from the mutant phenotype of a cell in which the peptide is altered or missing.

(2) By examining the phenotypes of cells in which normal and mutated genes are present in the same cell (complementation analysis), it may be possible to infer the minimum number of gene products required to mediate a biological process. We will discuss the experimental requirements for drawing each type of inference, with particular reference to the study of hormone action in S49 mouse lymphoma cells.

7.1.1 Mutant selection

It is necessary to begin with a permanent cell line that expresses one or more responses to a hormone in a stable fashion. Ideally, the cell line should be amenable to propagation by cloning individual cells, and all unselected clones should express similar or identical hormone responses. The S49 cell meets these criteria: It is a permanent or 'immortal' cell line, derived from a lymphoid tumor, it grows in suspension culture, exhibits a stable karyotype with 40 chromosomes (the diploid complement of mouse cells), and readily forms clones in soft agar which has been conditioned by a feeder layer of mouse fibroblasts; the cytocidal actions of cAMP and glucocorticoids are virtually identical in unselected wild-type clones.

Once these criteria are satisfied, the key to successful genetic analysis is to devise a method for separating mutant cells from the normal population of wild-type cells. In principle, this could be done simply by propagating each of a vast number of independent clones to a sufficient cell number for their phenotypes to be characterized, for example with respect to any measurable effect of a hormone on intact cells. Since mutant cells may appear at a frequency lower than one per million in an unselected wild type population, this approach is usually not feasible. Accordingly, it is necessary to devise an efficient selection procedure. The phenotype of wild-type S49 cells provides a ready-made selection procedure: since glucocorticoids and cAMP kill normal S49 cells, a clone that survives and multiplies in the presence of one of these effectors is a good candidate for characterization as a hormone-resistant mutant. In other cases it may be necessary to take advantage of some other hormone-induced difference between the normal cell population and putative mutants, such as differences in rate of growth, binding or production of an easily detectable marker substance, or uptake of a cytotoxic chemical.

When a variant clone has been selected by any of these procedures, it is crucial to determine whether its difference from wild-type cells is caused by gene mutation. If the variant phenotype is caused by some heritable alteration analogous to differentiation, rather than by mutation, the changes observed in the variant may not be causally related to one another. On the other hand, all the functional changes observed in a variant phenotype produced by the mutation of a single gene product are likely to be the result of that mutation.

Strictly speaking, absolute proof of mutation requires the demonstration of an altered sequence of bases in DNA. While this stringent criterion has not been met for any mutation selected in somatic cells grown in tissue culture, a variety of other criteria are used in practice to establish whether mutation is likely to be the cause of an altered phenotype. Roughly in order of increasing stringency, these criteria may be listed as follows:

(1) The altered phenotype should be heritable and stable in the absence of continued selective pressure.

(2) The altered phenotype should occur at a plausibly low frequency, generally comparable to mutation rates in well-defined microbial systems.

(3) The occurrence of variant clones should be random, and not dependent on selective pressure. This random occurrence of mutations can be tested by a procedure termed fluctuation analysis, described in detail elsewhere (Luria and Delbruck, 1943).

(4) Chemical mutagens should increase the incidence of variant clones.

(5) Demonstration of a change in the primary structure of a peptide gene product makes it highly likely that the phenotypic alteration is due to mutation.

No one of these criteria is sufficient (or absolutely necessary) to establish the existence of a mutation. According to these criteria, several of the S49 variant phenotypes which we will discuss are virtually certain to have resulted from a mutation; these will be termed mutant cells, while the term *variant* will be applied to heritable phenotypes which have not yet been shown to meet more than one or two of the criteria listed.

The distinction between mutant and phenotypically variant cells may bear on the strength of inferences we are able to draw about causal relations between functional alterations in a variant cell. The isolation of cells from a variant population that have reverted back to the wild-type phenotype may also constitute strong evidence in favor of such a causal relation. If the appearance of such revertants meets the same criteria applied to the original 'forward' mutation (random incidence, sensitivity to mutagens, etc.), all of the functional changes in the variant phenotype are quite likely to be related to the alteration of a single gene product.

It may be difficult to design an efficient selection in the 'backward' direction if 'forward' selection of the hormone-resistant phenotype depended on the ability of the hormone to kill wild-type cells. Perhaps for this reason, no revertants have yet been selected from S49 variants with altered responses to hormones. One potential method for isolating revertants, termed counter-selection, takes advantage of hormone-induced growth arrest of wild-type cells. An example of successful counterselection, used to isolate 'cAMP-deathless' S49 variants, is described below (Section 7.3.2).

7.1.2 Complementation analysis

In the initial stage of investigating a poorly understood biological response, it is useful to make some estimate of its biochemical complexity. In bacteria, genetic complementation analysis has often been used to obtain a minimum estimate of the number of distinct gene products involved in a particular cellular process. The principle is simple: firstly a series of independently selected mutants is obtained, each having lost a biochemical function performed by wild-type cells (e.g., bio-synthesis of an amino acid or response to a hormone). These mutants are then mated or 'crossed', and the daughter cells are examined for restoration of the wild-type phenotype. Crosses between mutants which bear lesions in different enzymes will complement one another to produce wild-type daughter cells, while crosses between mutants lacking the same enzyme will not. Providing that crosses between a large enough array of independent mutant cells are examined, the number of complement-ation groups (groups of mutants which pairwise do not complement one another) will provide a reliable minimum estimate of the number of gene products required for the normal wild-type function.

In somatic cells such crosses are made by the formation of cell—cell hybrids bearing chromosomes and genes from two parental cells. This requires a method for promoting cell fusion, and a procedure for separating hybrid cells from parental cells that have not fused. In S49 cells, fusion is promoted by incubation with inactivated Sendai virus or with polyethylene glycol. Hybrid cells are selected using the HAT (hypoxanthine-aminopterin-thymidine) selection system of Littlefield (1964). One parent must be deficient in hypoxanthine phosphoribosyl transferase (HPRT) activity and the other deficient in thymidine kinase (TK). S49 mutants of each type are readily selected by cloning the cells in 6-thioguanine or bromouridine-deoxyribose, respectively. HPRT-deficient cells cannot use exogenous purine sources such as hypoxanthine, and TK-deficient cells cannot use exogenous thymidine. Since aminopterin blocks endogenous synthesis of both purines and pyrimidines, cells incubated in HAT survive only if both HPRT and TK activities are present. When cells of each parental type fuse, the deficiency of each is complemented by the other, and the hybrid cells survive and form colonies in agar containing HAT, while both parental cell lines die.

A mutant phenotype – e.g., loss of response to a hormone – may be dominant or recessive in hybrids formed by fusion with wild-type. Dominance of the mutant phenotype (i.e., the hybrid is also hormone resistant) suggests that the altered phenotype is produced by action of diffusible gene products that are either *trans* inhibitors of allelic expression (i.e., able to regulate expression of active genes from another chromosome) or that modify the activity of other gene products. Mutant phenotypes that are recessive with respect to wild-type are more likely to result from deficiency of a particular gene product, and thus may be used to define complementation groups. We will discuss both dominant and recessive S49 phenotypes with lesions affecting hormone action, and some conclusions that can be drawn from hybrid phenotypes.

7.1.3 Search for the mutant gene product

For genetic analysis to be truly successful as a strategy for gaining understanding of a biological phenomenon such as hormone action, the peptide gene product directly altered by mutation must be unambiguously identified. In some cases this is a relatively easy task. For example, prior knowledge of a hormone's action may suggest that a lesion which affects a particular receptor or enzyme could produce the mutant phenotype. In several of these cases with S49 mutants, a lesion in such receptors has been sought and found, providing elegant proof that the receptor (or enzyme) does mediate the actions of the hormone.

In other cases, however, it may be difficult to define the biochemical lesion responsible for loss of hormone responsiveness in a variant cell. Ligand binding and enzymatic activities which are thought to mediate a hormone's action may appear intact in the variant. Such cases provide an opportunity and a challenge, because ultimate identification of the altered gene product is sure to increase our knowledge of the molecular basis of the hormone's action.

7.2 ACTION OF GLUCOCORTICOID HORMONES

Certain tissue-culture cells respond to treatment with glucocorticoids. In some cases, e.g., induction of liver-specific enzymes in hepatoma cultures, the hormonal response represents a close analog of regulatory processes that occur in the intact animal. Because of their relative amenability to experimental study, hormone-responsive cultured cells have contributed substantially to our current understanding of how steroid hormones work. In broad outline, the following processes are believed to occur. Steroid hormone crosses the cell membrane and enters the cytoplasm, probably by passive diffusion. It binds there to a high-affinity receptor, composed of two or more peptides. The interaction of hormone and receptor results in an 'activated' complex with altered physico-chemical properties, including

increased affinity for nuclear sites. The inactive receptor is predominantly localized
to the cytoplasm. Following activation, the receptors are transferred in substantial
degree to the nucleus. This nuclear binding results in specific interaction with
control sites on chromatin that alter the transcription rate of messages for specific
gene products.

The number of gene products responsive to glucocorticoids is limited (Ivarie
and O'Farrell, 1978) and the nature of the response is tissue-specific. The function
of lymphoid cells is generally inhibited by glucocorticoids. This accounts in part for
the anti-inflammatory and immunosuppressive effectiveness of these agents. In
murine species, some classes of lymphoid cells are killed by glucocorticoids, e.g.,
cortical thymocytes. An analogy of the cytolytic process is demonstrable in culture;
addition of natural or synthetic glucocorticoids to many lines of cultured mouse
lymphoid cells results in prompt cell death. This observation suggested a line of
investigation, carried out by Sibley, Yamamoto and Stampfer in Tomkins' lab,
based on the ability to isolate and study mutant cells with specific defects in their
response to the hormone.

The S49 mouse lymphoma carries a specific membrane antigen of T cells
(Hyman, 1973). Addition of 10^{-7} M dexamethasone, an artificial glucocorticoid,
to culture medium results in cell lysis that is complete within 24 h. The concentration-
dependence of this effect using a series of steroids exhibits an order of potency
characteristic of classic glucocorticoid responses (Rosenau *et al.*, 1972). When S49
cells are cultured with dexamethasone under conditions that allow single cells to
divide and form colonies, clones survive at a frequency of about 10^{-5}. These
surviving cells can be propagated and are found to be stably resistant to the selective
agent.

Several types of evidence contribute to the conclusion that the variant cells are
mutants. Their frequency is low. Fluctuation analysis indicates a mutation frequency
of 3.5×10^{-6} per cell per generation (Sibley and Tomkins, 1874a). The incidence
of mutant cells is enhanced by chemical mutagenesis (Sibley and Tomkins, 1974a).
The resistant phenotype persists indefinitely in the absence of the selective agent.
Perhaps most significantly, direct biochemical evidence demonstrates an alteration
of the target gene product, the glucocorticoid receptor protein.

The mutants show a variety of defects, but almost all affect the activity of the
cytoplasmic receptor (Sibley and Tomkins, 1974b). Three types have been
determined by measurement of binding and subcellular localization of radiolabelled
dexamethasone:

(1) Mutants with a profound deficiency in binding activity, termed receptor
minus (r^-).
(2) Mutants that fail to transfer steroid—receptor complex to the nucleus, but
have near-normal quantities of cytoplasmic binding activity, called nuclear transfer
minus (nt^-), and
(3) mutants that transfer greater than normal quantities of steroid—receptor

complex to the nucleus, termed nuclear transfer increased (nt^i) (Sibley and Tomkins, 1974b; Yamamoto *et al.,* 1974).

The low relative abundance of the receptor protein makes it technically unfeasible to determine whether the receptor is absent or functionally altered in r$^-$ mutants. The situation is quite different in the case of nt$^-$ and nt^i cells. In both cases, the receptor activity sediments differently from that of wild-type cells in sucrose gradients, suggesting an altered protein structure, composition or conformation (Yamamoto *et al.,* 1974). The affinities for DNA of nt$^-$ and nt^i are distinguishable from each other and from wild-type with nt^i > wild-type > nt$^-$. The relative binding affinity for DNA was determined by allowing the radiolabelled steroid−receptor complex to bind to a column of DNA-cellulose and then measuring the elution of the complex using a salt gradient. Receptor−steroid complexes of nt$^-$, wild-type and nt^i eluted respectively at NaCl concentrations of 110−120, 170, and 210 mM NaCl (Yamamoto *et al.,* 1974).

The defective nature of receptors in mutants which are unresponsive to gluco-corticoids strongly confirms the central role of that protein in the steroid response. The presence of functionally normal receptor is a necessary condition for the biological response to occur. Steroid transfer to the nucleus does not suffice; this happens in nt^i mutants, but does not result in an effective interaction. Binding of the receptor−steroid activated complex to DNA is likely to be an essential part of the cytoplasm-to-nucleus transfer process, because cytoplasmic−nuclear partitioning ratios among nt^i, wild-type, and nt$^-$ are predicted by their affinity for purified DNA.

Two related genetic questions may be posed. How is it possible to obtain these classes of receptor mutants in S49 cells? Why are mutants in non-receptor components of the response system seldom or never found among the hundreds of dexamethasone-resistant mutants screened? S49 cells have a near-normal karyotype, consisting of 40 acrocentric chromosomes (Coffino *et al.,* 1975). In a normal somatic cell, each genetic locus encoded on an autosome (non-sex chromosome) should be represented in two copies, one on each of two homologous chromosomes. In the simplest analysis, r$^-$ mutants should arise as a result of two events, each resulting in the alteration of receptor-coding capacity on a single chromosome. Worse, if nt$^-$ and nt^i mutants result from structural mutation at one of the homologous genes, the second should continue to code for normal receptor, and both classes of receptor should be detected in the mutant cells. This is not the case. We know that two receptor types can be simultaneously produced and detected in a single cell because this occurs in hybrid cells formed by fusing nt^i x wild-type (Yamamoto *et al.,* 1976). We must conclude that in wild-type S49 cells only one gene copy actively encodes the receptor, the activity of the other having been lost by an unknown mechanism, as has been postulated in other cultured cells.

Further evidence for this hypothesis comes from comparing S49 with the WEHI-7 lymphoma in studies carried out by Bourgeois and collaborators (Bourgeois and Newby, 1977; Bourgeois *et al.,* 1978). Attempts to select glucocorticoid-resistant variants of the latter cells were initially unsuccessful. Fully resistant cells could, however, be obtained in a two-step process, by first selecting for cells resistant to a relatively low concentration of drug and subsequently selecting from those cell variants resistant to a higher concentration of drug. Assays of the glucocorticoid-binding activity of each cell population revealed the following: the intermediate stage cells with low-level resistance had about half the number of receptors present in the parent cells, while the fully resistant cells had little or no binding activity. The number of receptors in S49 resembled that in low-level resistant WEHI-7. Furthermore, while parent WEHI-7 cells gave rise to fully resistant cells at a vanishingly low frequency (less than 1.6×10^{-9}) the ease of generating such fully resistant cells from partially resistant WEHI-7 clones and from S49 parental cells was similar. A plausible interpretation of this data is that WEHI-7 expresses two active genes for the receptor and that S49 expresses one. Gene dosage determines the relative synthetic rate for receptors and its steady-state level. A single-hit event, inactivating one gene, converts WEHI-7 to an S49-like state, genetically and phenotypically.

The questions posed above become answerable in terms of this analysis. Receptor mutants are readily obtainable in S49 because the target gene, presumably that coding for a structural component of the receptor, is present in only one functional copy. It is the target of opportunity. Perhaps there are other genes, controlling other elements of the response system, which would produce resistance upon mutation. If so, among S49 cells, they are not readily seen above the 'background' of receptor mutants. A solution which has been proposed (Bourgeois *et al.*, 1978) is to use WEHI-7, in which the noise level is much reduced, to select for non-receptor mutants. Selection on very large numbers of cells ($10^9 - 10^{10}$) would be required, but this is technically feasible.

Preliminary efforts to obtain new classes of mutants by this means have not yet succeeded, as they require consideration of this question: are post-receptor mutants merely rare or are they in principle non-existent? The latter could be the case if mutable post-receptor elements were essential for cell viability, if multiple redundant genes coded for identical or similar functions, or if no specific proteins were required for post-receptor events. Because of the importance of understanding the distal mechanisms of the glucocorticoid response and the difficulty of resolving these problems by purely biochemical means, the genetic approach will remain inviting.

7.3 cAMP ACTION

The capacity to make cAMP, to modulate its intracellular levels in response to external stimuli and to regulate the activity of protein kinases(s) in response to that level is a universal property of animal cells. Hormones and other effectors of cAMP appear to exert their biological effects as follows: firstly, increased levels of cAMP activate one or more protein kinases. cAMP-dependent kinase is tetrameric, consisting of two regulatory (R) and two catalytic (C) subunits. Binding of cAMP to the regulatory subunits results in dissociation of the holoenzyme to an R_2 dimer and two C subunits. The two C subunits are inactive in the holoenzyme and active in their free form. C catalyzes the phosphorylation of protein substrates, using ATP as a phosphate donor, and this results in alteration of the protein's enzymatic activity. The substrate specificity of C is exquisite; a limited number of proteins in a given cell are modified. Some of the known substrates include phosphorylase kinase, glycogen synthetase, pyruvate kinase and certain histones (Nimmo and Cohen, 1977).

To investigate the mechanisms of cAMP-mediated regulation in animal cells, S49 mutants with specific and defined defects in the response system were generated (Coffino *et al.*, 1975). Use was made of the fact that sustained elevation of cAMP, whether generated endogenously or by exogenous addition of an analog, kills S49 cells. Cloning of wild-type populations of cells in a medium which contains dibutyryl cAMP results in the survival of small numbers of colonies composed of cells resistant to killing by cAMP. The phenotype is stable in the absence of further selection. Fluctuation analysis reveals that resistant cells arise from sensitive parents at a rate of 2×10^{-7} per cell per generation (Coffino *et al.*, 1975). This frequency can be greatly enhanced by mutagenesis (Friedrich and Coffino, 1977). Cells selected to survive in dibutyryl cAMP are also resistant to another cAMP analog, 8-bromo-cAMP, and to agents such as β-adrenergic agonists and cholera toxin that kill wild-type S49 by stimulating cAMP synthesis.

Assay of the cAMP-dependent kinase activity of these cells reveals that they are almost uniformly defective in protein kinase activity (Bourne *et al.*, 1975a; Insel *et al.*, 1975). (An exception will be described below). Determination of the type and degree of deficiency in activity permits classification of the mutants into three categories. These are:

(1) Cells devoid of detectable cAMP-dependent kinase (kin^{-1}),

(2) Cells with kinase whose apparent affinity for cAMP is reduced compared to that of wild-type (K_a), and

(3) Cells with reduced amounts of kinase activity, whose apparent affinity for cAMP is similar to that of wild-type (V_{max}) (Insel *et al.*, 1975).

The degree of resistance to dibutyryl cAMP correlates directly with the kinase defect. Kin$^-$ cells are fully resistant at all concentrations, K_a mutants display toxicity at high concentrations, and V_{max} mutants exhibit reduced toxicity compared to wild-type, but the concentration of dibutyryl cAMP that elicits a half-maximal effect is similar in V_{max} mutants and wild-type cells (Insel *et al.*, 1975).

Other responses to cAMP found in wild-type S49 cells are also perturbed in the kinase deficient mutants. These include inhibition of growth in the G_1 phase of the growth cycle, induction of cAMP phosphodiesterase and the extinction of ornithine decarboxylase and S-adenosyl-methionine decarboxylase (Insel *et al.*, 1975; Insel and Fenno, 1978). When tested by addition of dibutyryl cAMP, each response is absent in kin$^-$, elicited at greater than normal concentrations in K_a, and reduced in magnitude in V_{max}. For all responses, there is excellent agreement between the nature of the enzyme deficiency determined by *in vitro* assay of its activity and the characteristics of the altered response of whole cells to the cAMP analog. These findings strongly support the central role of cAMP-dependent kinase in the action of cAMP in animal cells.

7.3.1 Molecular basis of kinase mutations

The molecular basis of the defects in the kinase mutants has been best determined in the K_a mutants. They have structural alterations in the R subunit. The change in cAMP-binding properties of the holoenzyme made this probable. Direct evidence came initially from studies in which R and C activity of wild-type and K_a mutant cells were fractionated and holoenzyme reconstituted homologously or heterologously (Hochman *et al.*, 1977). Properties of the mutant enzyme were conferred by fractions from mutant extracts containing R activity but not by those with C activity. Subsequently, two-dimensional polyacrylamide gel electrophoresis revealed that most K_a mutants have R subunits with an altered isoelectric point (Steinberg *et al.*, 1977).

Interestingly, K_a mutants, that by this criterion synthesize altered R, invariably make structurally normal R as well. Of two homologous loci coding for R, only one has undergone mutation. The cells make kinase that is a mixture of normal and altered molecules. Analysis by two-dimensional gels of K_a mutant cells pulse-labelled with ^{35}S-methionine demonstrates that the synthesis of mutant R subunits exceeds that of the normal peptide (Coffino and Gray, 1978). Regulation of kinase synthesis thus appears to exhibit *cis*-acting regulation, in the sense that synthetic rates of mutant and wild-type R are capable of separate control. The basis for this form of regulation remains to be examined. In any case, the amount of normal kinase (containing normal R subunits) is insufficient to confer on the cells a wild-type phenotype.

V_{max} mutants contain as much as half the kinase activity of wild-type cells, and this activity appears biochemically normal (i.e., in K_a for cAMP). These mutants are sufficiently resistant to cAMP to survive selection and they display a phenotype which shows reduced maximal responses to cAMP (Insel *et al.*, 1975). We can conclude from this not only that cAMP-dependent kinase is essential for cAMP action, but also that in wild-type cells the level of kinase activity is just sufficient to generate a maximal response to elevation of cAMP. If cAMP-stimulated protein phosphorylation mediates the effects of cAMP in S49 cells, the fact that the amount of cAMP-dependent kinase limits maximal effectiveness of cAMP may also indicate the existence of efficient biochemical mechanisms – e.g., specific protein phosphatases – for reversing these phosphorylations; otherwise, even a reduced amount of kinase would be able to mediate a maximal cAMP effect.

The mechanism responsible for the complete lack of cAMP-dependent protein kinase activity in kin⁻ mutants is not entirely clear, but the available evidence suggests that these cells have undergone mutation in a system responsible for regulation of the kinase. A striking property of these cells is that the mutant is fully *trans*-dominant: Hybrids formed by crossing wild-type x kin⁻ are kin⁻ (Steinberg *et al.*, 1978). Consideration of the several possibilities suggests that kin⁻ cannot be explained as a lesion of a known element in the kinase system. A heat-stable protein that binds to C has been described in several tissues and is present in S49 cells (Steinberg *et al.*, 1978). But kin⁻ has no more of this activity than wild-type. Further, in mixing experiments the cAMP-dependent kinase activities of extracts of kin⁻ and of wild-type display strict additivity. Hence the kin⁻ properties cannot result from the presence of an excess of a soluble inhibitor. The R subunit is made in kin⁻, although at a lower rate than in wild-type cells. On two-dimensional gels, the R of kin⁻ appears normal. It appears to be functionally normal as well, for addition of extracts of kin⁻ to partially purified C's from wild-type cells leads to asembly of cAMP-dependent kinase. Lastly, it is difficult to think of a plausible model ascribing kin⁻ to structural mutations of C, for such a model would have to explain both the dominance of kin⁻ and the observation that the R activity of kin⁻ fractionates in several chromatography systems as free R, not as holoenzyme (Steinberg *et al.*, 1978). By a process of elimination, we are led to the tentative conclusion that kin⁻ results from a lesion in a gene that controls a *trans*-dominant function, not previously described, which tightly regulates the expression of cAMP-dependent protein kinase.

Availability of kin⁻ mutants facilitates the analysis of the domain of cAMP-dependent protein phosphorylation in S49. A sensitive means for detecting protein phosphorylation is to analyse extracts of ^{35}S-methionine labelled cells by auto radiography of two-dimensional gels. The change in isoelectric point conferred by phosphorylation results in a readily perceptible shift of a peptide spot in the charge dimension. Phosphorylation mediated by cAMP-dependent kinase is detected as the partial or complete shift of a protein spot to a more acidic position in the gel when wild-type, but not the kin⁻, cells are treated to activate kinase, for example with hormones or congeners of cAMP. Analysis of this kind reveals that about 16 proteins undergo charge modifications consistent with phosphorylation when cAMP-dependent kinase is activated (Steinberg and Coffino, 1979). The high resolution of the gel system allows analysis of perhaps 1000 proteins in this manner. The substrate specificity of kinase is thus seen to be substantial. In addition to apparent protein modification, there are additional changes in the rate of labelling of individual spots that become apparent only after prolonged (two hours or longer) stimulation of kinase activity. The rate of synthesis of six proteins is increased, while synthesis of three proteins is decreased under these conditions. In kin⁻ cells, manipulations that in wild-type cells produce kinase activation have no effect on the pattern of protein labelling.

7.3.2 cAMP resistance with normal kinase

Among hundreds of S49 mutants selected for their ability to exibit clonal growth in the presence of dibutyryl cAMP, one has proven to be resistant due to a defect in a transport function, rather than in kinase activity (Steinberg and Steinberg, in press). This variant is resistant to killing and to growth inhibition by dibutyryl cAMP and by N^6-monobutyryl cAMP, but is sensitive to 8-bromo-cAMP and endogenous cAMP stimulated by cholera toxin. Accumulation of radiolabelled exogenous cAMP is reduced in the variant compared with wild-type, as is accumulation of endogenous cAMP after treatment with isoproterenol or cholera toxin. After treatment with isoproterenol, more cAMP appears in the extracellular medium of the variant cells than of the wild-type. They appear, therefore, to have an alteration that results in an increased rate of transport of cAMP and certain congeners out of the cells. The variant may prove of value in determining the mechanism and specificity of cAMP extrusion from cells.

One further class of mutants, termed cAMP deathless (cAMP D⁻) were selected to survive killing, but to retain the growth regulatory response to cAMP (Lemaire and Coffino, 1977). Wild-type cells are arrested in the G_1 phase of the growth cycle by cAMP. A mutagenized population of cells was exposed to dibutyryl cAMP for a period equivalent to two cell generations. During the last half of this time, cytosine arabinoside, a drug that kills cycling but not G_1-arrested cells, was added to the growth medium. The result of this counter-selection procedure was to kill wild-type cells by virtue of their sensitivity to dibutyryl cAMP and to kill kinase-deficient cells with cytosine arabinoside because their progression through the cycle was not inhibited. Cloning the resultant survivors yielded cells with the cAMP D⁻ phenotype. These cells have normal kinase activity and biological responses to cAMP, except for killing, that are quantitatively identical to those of wild-type cells. The cAMP D⁻ variant has proven useful in studies of growth regulation by cAMP (Coffino and Gray, 1978) and may help in establishing the mechanism of S49 cell killing by cAMP.

7.4 HORMONE-SENSITIVE ADENYLATE CYCLASE

Animal cell membranes transduce the information provided by an external hormonal signal into synthesis of cAMP, the intracellular 'second messenger'. Hormone receptors and adenylate cyclase are present at low concentrations in cell membranes, and appear functionally labile when separated from the lipid membrane milieu. As a result, it has become possible only recently to ask simple questions regarding the number and nature of distinct molecular species that comprise hormone-sensitive adenylate cyclase. Biochemical investigation of variant S49 clones has contributed significantly to recent advances in our understanding.

Because β-adrenergic amines, prostaglandin (PG) E_1 and cholera toxin stimulate cAMP synthesis in S49 cells, these effectors can be used as selective agents for isolating genetic variants with lesions affecting hormone-sensitive adenylate cyclase. Three classes of such variants have been isolated (Table 7.1). Historically, the first variant was selected in agar which contained isoproterenol and the phosphodiesterase inhibitor RO 20–1724 (Bourne *et al.*, 1975b). Adenylate cyclase in this variant

Table 7.1 S49 variants with lesions affecting hormone-sensitive adenylate cyclase

Clone	Wild-type	cyc^-	UNC	β_d
Stimulation of cyclase by:				
isoproterenol	+	−	−	↓↓
PGE_1	+	−	−	+
Gpp(NH)p	+	−	+	+
cholera toxin + NAD^+	+	−	+	+
NaF	+	−	+	+
Mn^{2+}	+	+	+	+
β-receptors*	+	+	+	↓↓

* Determined by binding of ^{125}I-hydroxybenzylpindolol. Wild-type cells possess 1000–1200 binding sites per cell, or about 100 fmol per mg protein in partially purified plasma membranes. All the variants possess similar numbers of β-adrenergic receptors, except for β_d, in which the number of receptors is decreased by 75–90% (Johnson *et al.*, 1979).

failed to respond to five different effectors that stimulate wild-type S49 cyclase: Catecholamines, PGE_1, cholera toxin, guanyl nucleotides, and NaF. The simplest interpretation was that this variant functionally lacked catalytic adenylate cyclase, and it was therefore designated AC⁻ (presently referred to as cyc^-). However, cyc^- cells were shown to contain cAMP although it was present at a concentration lower than the basal level measured in wild-type (Bourne *et al.*, 1975b). These variant cells also possess the normal (wild-type) complement of β-adrenergic receptors, as determined by radioligand binding (Insel *et al.*, 1976). Subsequently, it was demonstrated that membranes of cyc^- cells possess Mn^{2+}-stimulated adenylate cyclase activity similar to that found in membranes of wild-type (Ross *et al.*, 1978; Vigne *et al.*, 1978). These and other results (Ross and Gilman, 1977a and b; Johnson *et al.*, 1978a and b) suggest that the cyc^- phenotype may result, not from a loss of catalytic adenylate cyclase, but rather from loss of a component required for its activation by hormones, cholera toxin, guanyl nucleotides, and NaF.

Using a similar selection procedure, Haga *et al.* (1977) isolated a second S49 variant, which they termed 'uncoupled' (UNC) (Haga *et al.*, 1976b). The UNC variant possesses a normal (or increased) number of β-adrenergic receptors and an adenylate cyclase stimulated by cholera toxin, guanyl nucleotides, and NaF. However, UNC cells and membranes respond poorly or not at all to both catecholamines and PGE_1. Thus it appears that the UNC phenotype is caused by a lesion affecting the mechanism that couples hormone receptor to cyclase activation.

A third class of S49 membrane variants is deficient in β-adrenergic receptors (Johnson *et al.*, 1979). Adenylate cyclase in these β_d variants responds normally to PGE_1 and to most other effectors, but shows specifically diminished responsiveness to β-adrenergic amines. In these variants the number of β-adrenergic receptors (measured by radioligand) is diminished in direct relation to the decrease in catecholamine responsiveness. Some of the variants show up to 90% loss in adrenergic binding sites and 90% loss in the ability of catecholamines to stimulate adenylate cyclase activity. The simplest explanation is that the β_d cells possess fewer β-adrenergic receptors than wild-type S49.

Examination of these three phenotypes provides considerable information about the complexity of hormone-sensitive adenylate cyclase, and allows certain otherwise attractive hypotheses to be ruled out. For example, in wild-type S49 and many other cell types, much lower concentrations of a β-adrenergic agonist (e.g., isoproterenol) are required to maximally activate adenylate cyclase than to saturate receptor binding sites. One proposed explanation of this discrepancy is that a large fraction of the β-adrenergic receptors are 'spare', in that they are not necessary for maximal stimulation of the cell's adenylate cyclase. The β_d phenotype rules out this simple explanation of the discrepancy, since a decrease in numbers of detectable β-adrenergic binding sites is associated with a directly proportional decrease in maximal stimulation of adenylate cyclase by β-adrenergic agonists (Johnson *et al.*, 1979). An alternative explanation will be described below (Section 7.4.3).

These phenotypes confirm and extend conclusions drawn from other kinds of experiments that hormone-sensitive adenylate cyclase is composed of more than one class of component molecule. The presence of β-adrenergic receptors in *cyc⁻* cells suggested that β-adrenergic receptors and cyclase were products of separate genes (Insel *et al.*, 1976). At about the same time, evidence was accumulating that hormone receptors and catalytic adenylate cyclase activity could be physically separated by gel filtration (Limbird and Lefkowitz, 1977), sucrose density gradient centrifugation (Haga *et al.*, 1977a) and affinity chromatography (Vauquelin *et al.*, 1977), and that β-adrenergic receptors of one cell type could be coupled to stimulation of adenylate cyclase provided by another cell, if the two were allowed to interact in the membrane of heterokaryons (Orly and Schramm, 1976; Schramm *et al.*, 1977). These experimental findings support the hypothesis that receptors are mobile in the plane of the membrane (Singer and Nicolson, 1972; Bennet *et al.*, 1975) and are capable of diffusing and interacting with other components involved in coupling receptors to cAMP synthesis.

7.4.1 Guanyl nucleotides and receptor-cyclase coupling

If receptors and catalytic adenylate cyclase are separate proteins, how does the interaction of a hormone with its receptor activate adenylate cyclase? Considerable insight into the possible mechanism derives from the significant observation that GTP markedly enhances the activation of adenylate cyclase by hormones in partially purified membrane preparations (Rodbell *et al.*, 1971a and 1975). In fact, in partially purified membranes from S49 cells, stimulation of adenylate cyclase by hormone shows an absolute requirement for the presence of a purine triphosphate (GTP or ITP) (Ross *et al.*, 1977). In addition, guanyl-5'-imidodiphosphate (Gpp(NH)p), a GTP analog whose terminal phosphate is resistant to hydrolysis by cellular GTPases, is capable of markedly activating adenylate cyclase by itself whereas GTP alone has little or no effect (Londos *et al.*, 1974; Schramm and Rodbell, 1975). Hormones that stimulate adenylate cyclase accelerate the rate of activation by Gpp(NH)p but not the maximal activity achieved (Ross *et al.*, 1977). Activation of adenylate cyclase by Gpp(NH)p or other hydrolysis-resistant GTP analogs is quasi-irreversible even with thorough washing of the membranes (Ross *et al.*, 1977; Schramm and Rodbell, 1975). This is in contrast to activation of adenylate cyclase by hormone and GTP, which is readily reversible by washing or addition of a hormone antagonist to the reaction mixture (Ross *et al.*, 1977; Schramm and Rodbell, 1975). These findings have led to the hypothesis that coupling of hormone receptors with activation of adenylate cyclase is controlled by guanyl nucleotide regulatory components of the adenylate cyclase system (Rodbell *et al.*, 1975; Maguire *et al.*, 1977).

A model for the regulation of adenylate cyclase activity by guanine nucleotides has been proposed by Cassel and Selinger (1977), based on their studies of the catecholamine-sensitive adenylate cyclase system of turkey erythrocyte membranes (Cassel and Selinger, 1976 and 1977). They propose a two-state model for adenylate cyclase, in which the GTP-liganded form of the enzyme is active; hormone increases the rate of formation of this GTP-liganded state (Cassel and Selinger, 1977). They have demonstrated an isoproterenol-stimulated GTPase activity in the turkey erythrocyte membrane, and propose that the GTPase returns the activated state of the enzyme to the basal inactive state by hydrolyzing the bound GTP to GDP + Pi. In this model, GDP-liganded adenylate cyclase is inactive, but in the continued presence of hormone, GDP is replaced by GTP, activating the enzyme. Gpp(NH)p activates adenylate cyclase in a quasi-irreversible fashion because of its hydrolysis-resistant terminal phosphate, which results in accumulation of the enzyme in the activated guanyl triphosphate-liganded state. Although hormone-stimulated GTPase activity has not yet been reported in other membranes, the model is generally consistent with studies of guanyl nucleotide regulation of adenylate cyclase activity in a number of different systems.

Guanyl nucloetides also affect the binding of agonists to specific hormone receptors (Ross *et al.*, 1977; Maguire *et al.*, 1977; Rodbell *et al.*, 1971b). GTP, GDP, and Gpp(NH)p shift the K_d for agonist binding to β-adrenergic receptors to higher concentrations, with no effect on antagonist binding (Maguire *et al.*, 1976). In the glucagon-sensitive adenylate cyclase system in hepatic membranes, GTP and Gpp(NH)p also decrease the affinity of the glucagon receptor for the hormone (Rodbell *et al.*, 1971; Lin *et al.*, 1977). Whereas Gpp(NH)p-activation of adenylate cyclase is quasi-irreversible, the effects of Gpp(NH)p on hormone binding are rapidly reversed by dilution (Ross *et al.*, 1977). These findings have led Rodbell and co-workers to postulate that functionally (and possibly structurally) distinct guanyl nucleotide regulatory sites are involved in the regulation of receptor binding and adenylate cyclase activity (Lad *et al.*, 1977).

How do guanyl nucleotides regulate hormone binding and adenylate cyclase activity in β_d, UNC, and *cyc⁻* variants? Activation of adenylate cyclase by Gpp(NH)p in membranes of the β_d variant is similar to wild type. GTP also shifts the K_d for binding of isoproterenol to higher concentrations for both the β_d variant and wild-type. These results indicate that the guanyl nucleotide regulatory sites of the β_d phenotype are similar to wild-type. Adenylate cyclase in membranes from UNC cells is activated by Gpp(NH)p, although the activation is somewhat reduced from that observed in wild-type membranes. Isoproterenol and PGE₁ in the presence of GTP have essentially no effect on UNC adenylate cyclase activity, whereas in wild type membranes they cause approximately 10-fold activation of the enzyme. Haga *et al.* (1977b) found a small effect of isoproterenol on the rate of activation of adenylate cyclase activity in membranes from UNC cells. The effect of guanyl nucleotides on the K_d for binding of agonists to the β-adrenergic receptor is lost in membranes from UNC cells. Isoproterenol binds to UNC receptors with the same affinity in the presence or absence of GTP; this affinity is identical to that observed with wild-type β-receptors in the presence of GTP (Ross *et al.*, 1977). Thus, the hormone receptor behaves as if it is unable to interact with the guanyl nucleotide regulatory sites and adenylate cyclase. Membranes from *cyc⁻* cells show no measurable adenylate cyclase activation or shift in K_d for agonist binding in the presence of GTP or Gpp(NH)p. The affinity of isoproterenol for the β-adrenergic receptor in *cyc⁻* membranes is also reduced, as in UNC, and is similar to the affinity of isoproterenol for the wild-type receptor in the presence of GTP (Ross *et al.*, 1977). Thus, it appears that the *cyc⁻* phenotype is completely deficient in guanyl nucleotide regulation of adenylate cyclase activity and of hormone-receptor binding.

Complementation analysis and coupling of receptors to cyclase

The UNC and *cyc⁻* phenotypes are similar in that both have lost guanyl nucleotide regulation of receptor affinity for agonists, and neither show stimulation of adenylate cyclase by hormones. The lesions differ, however, in that the UNC cyclase can be stimulated by several non-hormonal effectors, including Gpp(NH)p. Hybrid clones constructed by crossing UNC or *cyc⁻* with wild-type S49 cells showed a wild-type phenotype, indicating that neither lesion was produced by a diffusible intracellular inhibitor of adenylate cyclase, and suggesting instead that each variant lacked one or more components of the cyclase system present in wild-type (Naya-Vigne *et al.*, 1978). The difference between the UNC and *cyc⁻* phenotypes suggested that they should be deficient in different cyclase components, and that therefore the UNC x *cyc* hybrid should show a wild-type phenotype (see Section 7.1.1). Surprisingly, UNC and *cyc⁻* lesions fail to complement one another, either in viable somatic cell hybrids (Naya-Vigne *et al.*, 1978) or in heterokaryons (Schwarzmeier and Gilman, 1977): In both cases the result of the combination resembles UNC.

These results suggest that the UNC and *cyc⁻* phenotypes share a common lesion, which may affect a specific receptor—cyclase coupling factor. An additional lesion must be postulated in *cyc⁻* to account for its differences from UNC. This additional lesion does not appear to be loss of catalytic adenylate cyclase itself, for reasons described above and including the demonstration of Mn^{2+}-stimulated cAMP synthesis by *cyc⁻* membranes (Ross *et al.*, 1978; Naya-Vigne *et al.*, 1978). Thus the S49 variant phenotypes suggest that hormone–sensitive adenylate cyclase may be composed of a minimum of four distinct components: receptors, catalytic adenylate cyclase, a coupling factor (missing in both UNC and *cyc⁻*), and a regulatory component lacking in *cyc⁻*. To reconcile this interpretation with the failure of complementation between UNC and *cyc⁻*, it is necessary to imagine coordinate loss of two membrane components in *cyc⁻*, produced by some mechanism that does not interfere with expression of wild-type components in wild-type x *cyc⁻* hybrids (which show the wild-type phenotype).

Alternatively, hormone-sensitive adenylate cyclase could be composed of a minimum of three components: receptors, catalytic cyclase, and a single class of coupling molecules that are probably involved in regulation of cyclase by guanyl nucleotides. The coupling molecule is absent in *cyc⁻*, and the UNC phenotype is produced by an alteration of the coupling molecule or by a reduction of its concentration in membranes. Such a change in the coupling molecule could impair its interaction with hormone receptors, but allow it to mediate cyclase activation by guanyl nucleotide, cholera toxin, and NaF.

7.4.2 Combining genetic and biochemical investigation

Complementation analysis provides strong hints as to the number of components that comprise hormone-sensitive adenylate cyclase. This approach, however, is subject to limitations of interpretation and requires a series of variants affecting different components. Ultimately, one must separate the components, purify them to homogeneity, and recombine them under defined conditions to produce a normally functioning system. This has proved an arduous task with a number of membrane-associated enzyme systems, including the Ca^{2+}–ATPase of sarcoplasmic reticulum (le Maire *et al.*, 1976a and b; le Maire *et al.*, 1978), Mg^{2+}–ATPase systems from bacteria (Vogel and Steinhart, 1976; Sone *et al.*, 1978; Abrams and Smith, 1974) and eucaryotic cells (Senior, 1973), Na^{+}–Ka^{+}–ATPase from kidney renal medulla (Goldin and Tong, 1974; Kyte, 1971; Goldin, 1977) and the oxidative phosphorylation system of mitchondria (Racker, 1970). The problems confronting resolution, purification, and reconstitution of hormone-sensitive adenylate cyclase are particularly great. Unlike the enzyme systems cited above, adenylate cyclase occurs in extremely low abundance in the plasma membranes of all cell types so far examined. The concentration of β-adrenergic receptors in S49 cells, for example, is approximately $100-150$ fmol mg^{-1} membrane protein or about 1000 receptors per cell (Insel *et al.*, 1976). This would mean that β-adrenergic receptors would comprise less than 0.0001% of total cellular protein. The concentrations of the regulatory components and catalytic adenylate cyclase probably do not differ vastly from that of the β-adrenergic receptor. Thus, classical biochemical techniques will have to be combined with a great deal of ingenuity to purify each component of hormone-sensitive adenylate cyclase.

(a) *Reconstitution experiments*

Genetic variants have provided a useful starting point in biochemical analysis of adenylate cyclase components. Reasoning that the missing component(s) of cyc^- might be present in extracts of normal cell membranes, Ross and Gilman first attempted biochemical reconstitution of adenylate cyclase by mixing detergent extracts of wild-type S49 membranes with intact membranes partially purified from cyc^-: Surprisingly, a hormone-stimulated adenylate cyclase was reassembled (Ross and Gilman, 1977a). It was shown that the cyc^- membranes were not simply acting as a membrane surface for the interaction of wild-type adenylate cyclase components, and that components from both the cyc^- membranes and the wild-type detergent extracts were required for enzyme activity (Ross and Gilman, 1977a and b). In addition to reconstitution of hormone sensitivity, the reassembled system was stimulated by Gpp(NH)p and by NaF. Using N-ethylmaleimide (NEM) or heating of the detergent extract it was possible to inactivate virtually all the adenylate cyclase activity in the extract without inhibiting the extract's ability to reconstitute a hormone-sensitive adenylate cyclase. The component(s) in the detergent extract is sensitive to proteases, suggesting that it is a protein. Inactivation of the Mn^{2+}-sensitive adenylate cyclase activity of cyc^- by NEM or heat also destroys the ability of cyc^- to reconstitute hormone-sensitive cyclase activity when mixed with wild-type extracts, suggesting that the cyc^- membranes contribute catalytic adenylate cyclase to the reconstituted system.

Interestingly, addition of detergent extracts from UNC to cyc^- membranes reconstitutes an adenylate cyclase stimulated by Gpp(NH)p and NaF, as would be predicted by the responses of membranes from UNC cells to these effectors; however, no hormone stimulation is observed. This is consistent with the cell hybridization experiments in which UNC and cyc^- failed to complement one another to give the wild-type phenotype.

Evidence that the component(s) donated by the detergent extract in the reconstituted system with cyc^- membranes are regulatory proteins, involved in guanyl nucleotide and fluoride regulation of adenylate cyclase, was obtained from experiments with different cell types deficient in specific hormone receptors or catalytic adenylate cyclase activity. B82 mouse L cells have a PGE_1-responsive adenylate cyclase but lack β-adrenergic receptors as measured by radioligand binding studies and the ability of catecholamines to stimulate adenylate cyclase (Maguire *et al.*, 1976; Brunton *et al.*, 1977). Detergent extracts from membranes of B82 cells, when mixed with cyc^- membranes, reconstitute a catecholamine-stimulated adenylate cyclase similar to that observed with heat- or NEM-treated detergent extracts from membranes of wild-type S49 cells. In addition, membranes of HTC cells possess β-adrenergic receptors and exhibit a guanyl nucleotide-induced shift in agonist binding to the β-receptor, but have virtually no stimulable adenylate cyclase activity. Detergent extracts from membranes of HTC cells, however, are capable of reconstituting a catecholamine-sensitive adenylate cyclase when mixed with cyc^- membranes. Since cyc^- membranes possess β-adrenergic receptors and a catalytic adenylate cyclase (stimulated by Mn^{2+}) which is similar to that observed in extracts from membranes of wild-type S49 cells, the simplest interpretation of these findings is that the cyc^- membranes donate the catalytic adenylate cyclase and the β-adrenergic receptor, while the detergent extracts of normal membranes donate the regulatory component(s) required for the actions of guanyl nucleotides and fluoride ion.

In summary, the cyc^- lesion appears to have accomplished by mutation a separation procedure that would be difficult or impossible using classical biochemical techniques: this lesion has specifically removed (or inactivated) a component (or components) of hormone-sensitive adenylate cyclase, leaving hormone receptors and catalytic cyclase functionally intact, but uncoupled, and not susceptible to regulation by effectors such as guanyl nucleotides. Taking advantage of this genetic lesion, the reconstitution procedure devised by Ross and Gilman opens the way to investigation of the mode of interaction of receptors and cyclase with the factor(s) missing in cyc^-. In addition, the reconstitution procedure provides a sensitive assay that may prove useful in purifying and characterizing components missing in both cyc^- and UNC membranes.

(b) *Cholera toxin as a tool*

Because it provides an alternative method for identifying one of the components of the cyclase system, cholera toxin can serve as a useful tool, complementary to the reconstitution procedure, for elucidating the molecular basis of hormonal stimulation of cAMP synthesis. Cholera toxin, a protein excreted by *Vibrio cholerae,* activates adenylate cyclase in intestinal epithelium. The resulting cAMP-stimulated increased transport of electrolytes into the gut lumen, accompanied by the passive flow of water, produces the massive fluid loss characteristic of cholera. Because cholera toxin stimulates adenylate cyclase in virtually every animal cell or tissue exposed to it, it is likely that the toxin acts on a component common to most membrane-bound cyclase systems.

In contrast to the almost instantaneous activation of adenylate cyclase by hormones, cholera toxin's stimulation of cAMP synthesis is preceded by a characteristic lag phase of 15–30 min. This lag phase is probably attributable to the binding of B subunits of the toxin to ganglioside GM_1 in the plasma membrane and the subsequent introduction of the toxin's active A subunit into the cell; the A subunit then modifies a component of adenylate cyclase (Gill, 1977). Once accomplished, activation of adenylate cyclase by cholera toxin is essentially irreversible, again in contrast to the stimulatory effects of hormones, which are readily reversed by dilution or addition of a specific hormone antagonist. Cholera toxin also activates adenylate cyclase in purified membrane preparations, but only in the presence of NAD^+ (Gill and King, 1975; Gill, 1975; Johnson and Bourne, 1977). Because the toxin's A subunit can act directly on adenylate cyclase, the characteristic lag phase observed with intact cells is absent.

Recent evidence suggests that cholera toxin activates adenylate cyclase by modifying its guanyl nucleotide regulatory component(s). Cholera toxin treatment of plasma membranes from wild-type S49 (Johnson and Bourne, 1977) and other mammalian cell types (Bitensky *et al.*, 1975), as well as those of avian erythrocytes (Cassel and Selinger, 1977), markedly increases the sensitivity of adenylate cyclase to stimulation by GTP, but not to stimulation by Gpp(NH)p or other hydrolysis-resistant GTP analogs. In turkey erythrocyte membranes, cholera toxin treatment inhibits an isoproterenol-stimulated GTPase activity (Cassel and Selinger, 1977); according to the model of Cassel and Selinger (1977) described above (Section 7.4.1), this inhibition of GTPase enhances the action of GTP, making it resemble the action of Gpp(NH)p.

Cholera toxin has been shown to possess NADase activity (Moss *et al.*, 1976) and to catalyze the transfer of the ADP-ribose moiety of NAD$^+$ to arginine (Moss and Vaughn, 1977). Taken together with the requirement for NAD$^+$ to activate adenylate cyclase, these observations suggest that the toxin acts enzymatically (like diphtheria toxin; Honjo *et al.*, 1969), by catalyzing ADP-ribosylation of a protein; this protein is probably a component of adenylate cyclase. Thus it seemed reasonable to ask whether the putative substrate of cholera toxin was present in wild-type and variant S49 membranes. When mixed with *cyc*$^-$ membranes, detergent extracts of cholera toxin-treated wild-type S49 membranes yield an adenylate cyclase system that is as responsive to GTP as to Gpp(NH)p; in the absence of cholera toxin treatment, however GTP has little or no effect (Johnson *et al.*, 1978a). This finding indicates that the substrate for cholera toxin is one of the components donated by the detergent extract, and suggests that it is involved in guanyl nucleotide regulation of the reconstituted enzyme.

If cholera toxin's substrate or site of action is normal in *cyc*$^-$ membranes, then toxin treatment of *cyc*$^-$ should produce a toxin-activated adenylate cyclase when these variant membranes are reconstituted by mixing with extracts from normal membranes. This toxin effect would be observed as an increase in adenylate cyclase activity measured in the presence of GTP, relative to that in the presence of Gpp(NH)p; toxin characteristically increases the GTP:Gpp(NH)p stimulation ratio. Conversely, if the cholera toxin substrate is altered or absent in *cyc*$^-$, cholera toxin treatment of *cyc*$^-$ should not affect the GTP:Gpp(NH)p stimulation ratio measured in the reconstituted system. Experiments showed that cholera toxin's effect could be seen in the reconstituted system only if the wild-type component of the mixture had been previously treated with toxin; enhanced GTP sensitivity was not seen when only the *cyc*$^-$ component had been exposed to toxin (Johnson *et al.*, 1978a). These experiments indicated that the cholera toxin substrate is functionally deficient in *cyc*$^-$.

If cholera toxin acts by catalyzing specific ADP-ribosylation, it should be possible to label the toxin substrate(s) with ^{32}P–NAD$^+$. When wild-type S49 membranes are incubated with ^{32}P–NAD$^+$, in the presence or absence of cholera toxin, a large number of membrane proteins is nonspecifically labelled, as determined by

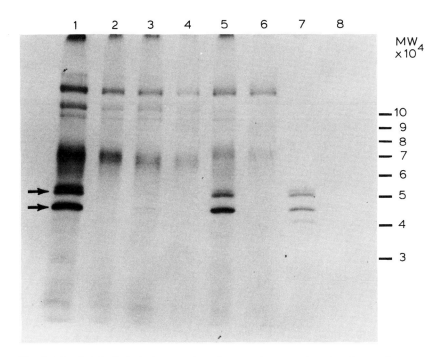

Fig. 7.1 Radiolabelled cholera toxin substrates in normal and variant cells. Each lane represents an autoradiogram of detergent extracts of membranes exposed to $^{32}P-NAD^+$ in the presence or absence of cholera toxin, and then subjected to polyacrylamide gel electrophoresis in the presence of sodium dodecyl sulfate: (1) cholera toxin-treated and (2) control wild-type S49 membranes; (3) cholera toxin-treated and (4) control cyc^- membranes; (5) cholera toxin-treated and (6) control UNC membranes; (7) cholera toxin-treated and (8) control membranes of HTC$_4$ cells. Taken, with permission of *The Journal of Biological Chemistry,* from Johnson *et al.,* 1978b.

polyacrylamide gel electrophoresis in sodium dodecyl sulfate. However, at least two prominent protein bands are specifically labelled in the presence of cholera toxin, with apparent molecular weights of approximately 45 000 and 52 000– 53 000 for the doublet of higher molecular weight (Johnson *et al.,* 1978b) (Fig. 7.1). If intact wild-type S49 cells are incubated with cholera toxin prior to the preparation of membranes, the labelling of toxin-specific bands is prevented, whereas the nonspecific labelling is unchanged or even enhanced. This finding strongly suggests that the 45 000 and 52 000–53 000 bands in purified plasma membranes represent the relevant toxin substrates in the intact cell.

Both the toxin-specific labeled bands are absent or greatly reduced in membranes of cyc^- cells (Johnson *et al.,* 1978b) (Fig. 7.1). This finding is consistent with the functional absence of cholera toxin substrates in cyc^-, as demonstrated in reconstitution experiments (Johnson *et al.,* 1978a). In contrast, specific peptides labelled by incubation with cholera toxin and $^{32}P-NAD^+$ are present in membranes of at least four other cell types, each of which also contains the detergent-soluble factor(s) which can reconstitute hormone-sensitive adenylate cyclase activity when mixed with *cyc*⁻ membranes. Membranes of HTC cells possess toxin-specific bands of Mr = 45 000 and 52 000—53 000 just as in wild-type S49 (Johnson *et al.,* 1978b). Erythrocytes of three species (man, Kaslow *et al.,* 1979; turkey, Kaslow *et al.,* 1979; and pigeon, Gill and Meren, 1978; Cassel and Pfeuffer, 1978) exhibit the toxin-labelled band of lower molecular weight, but appear to lack completely the higher molecular weight doublet. Interestingly, membranes of two of these cell types, the human erythrocyte and the HTC cell, are virtually devoid of catalytic adenylate cyclase activity, even in the presence of Mn^+ ion (Ross *et al.,* 1978; Kaslow *et al.,* 1979).

Taken together, all these findings indicate that the membrane substrate(s) of cholera toxin is identical to, or closely associated with, the regulatory factor(s) absent in *cyc*⁻ membranes that is required for coupling of hormone receptors to stimulation of cAMP synthesis and that mediates regulation of adenylate cyclase by guanyl nucleotides and NaF. The toxin's ability to elevate the GTP:Gpp(NH)p stimulation ratio suggested that the toxin substrate bears the guanyl nucleotide regulatory site of adenylate cyclase. This conclusion is greatly strengthened by recent observations that the toxin-labeled substrate of pigeon erythrocyte membranes binds specifically to a GTP affinity column (Cassel and Pfeuffer, 1978), and that a similar column specifically retains the factor in detergent extracts of human erythrocyte membranes that reconstitute hormone-sensitive adenylate cyclase when mixed with *cyc*⁻ membranes (unpublished result). Such columns may provide the first step in purification of the guanyl nucleotide regulatory protein(s).

The UNC phenotype cannot be explained simply by the loss of one of the major toxin substrates, since both the toxin-labeled bands present in wild-type are also labelled in UNC membranes (Johnson *et al.,* 1978b) (Fig. 7.1). However, this finding does not rule out the possibility of a structural alteration in one or both of the toxin-specific substrates of UNC that makes it unable to couple hormone receptors with adenylate cyclase, but preserves both its capacity to serve as a substrate for cholera toxin and its ability to mediate guanyl nucleotide and NaF regulation of cAMP synthesis. Alternatively, as described earlier, it is possible that the UNC membranes lack an additional protein, also missing in *cyc*⁻, that is not a toxin substrate.

7.4.3 Proposed model for hormone-sensitive adenylate cyclase

On the basis of the experimental findings reviewed in this section, we have proposed a tentative model that accounts for regulation of adenylate cyclase by guanyl nucleotides and hormones (Johnson *et al.,* 1978a). The model extends that proposed by Cassel and Selinger (1977), described in Section 7.4.1, above. Hormone-sensitive

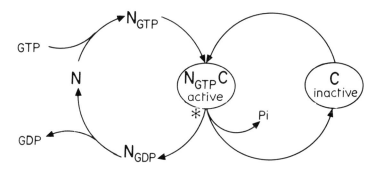

※ Site of toxin action

Fig. 7.2 Proposed model for guanine nucleotide regulation of adenylate cyclase activity. See text for details. Taken, with permission of *The Proceedings of the National Academy of Sciences*, from Johnson *et al.*, 1978a.

adenylate cyclase is envisioned to be composed of three distinct components: Hormone receptors (R), catalytic adenylate cyclase (C), and a regulatory-coupling element which we have termed N because of its involvement in regulation by guanyl nucleotides. N and C are considered to be separate from one another in the membrane, unless N binds a guanyl triphosphate (Fig. 7.2). The N_{GTP} complex binds tightly to C, and the ternary complex of N, GTP, and C actively synthesizes cAMP from ATP. The ternary complex, however, also possesses GTPase activity: when the bound GTP is hydrolyzed, the complex dissociates, releasing free (catalytically inactive) C and N_{GDP}. A new N_{GTP}C ternary complex can only be formed if GDP is released from N and replaced by GTP, a process that is accelerated by hormone-receptor (HR) complex, which may act by promoting the release of GDP (Cassel and Selinger, 1978).

In this model the proportion of C in the active, cAMP-synthesizing state (the ternary complex with N and GTP) is determined by two rates:

(1) the rate at which N_{GTP} is generated from N (or from N_{GDP}), a process regulated by HR;

(2) the rate of hydrolysis of GTP in the ternary complex.

The latter rate is slow when Gpp(NH)p is substituted for GTP, leading to greater overall activation of cAMP synthesis. In addition, ADP-ribosylation of N by cholera toxin prevents or slows hydrolysis of GTP, similarly leading to greater activation of cyclase by GTP. Note that this effect of cholera toxin is closely analogous to the well documented effect of diphtheria toxin's ADP-ribosylation of Elongation Factor 2 (EF-2), which blocks GTP hydrolysis by a complex of EF-2, ribosome, mRNA, and aminoacyl-tRNA (Moss *et al.*, 1976, Raeburn *et al.*, 1968; Skogerson and Moldave, 1968).

While the proposed model is not uniquely determined by available data, it is consistent both with kinetic studies of adenylate cyclase activation by hormones and guanyl nucleotides and with the variant S49 phenotypes. It should prove useful as a framework for exploring a variety of questions related to hormone action in membranes:

(1) More complete understanding of hormone-sensitive adenylate cyclase will come from purification and characterization of its components, a goal actively pursued in many laboratories. As these components become better defined in molecular terms, it will become possible to test implications of the model for their mode of interaction in membranes.

(2) The model's focus on N as an intermediary between receptors and activation of cyclase raises the possibility that N (or similar GTP-binding molecules) mediates hormonal regulation of other membrane enzymes. In this regard it is interesting to note that guanyl nucleotides have been shown to regulate binding to membrane receptors of several classes of hormones and effectors that do not work by stimulating cAMP synthesis, including morphine and endogenous opiate-like molecules (Blume, 1978; Childers and Snyder, 1978), α-adrenergic agonists (U'Prichard and Snyder, 1978), and angiotensin (Glossman *et al.,* 1974).

(3) By interposing N_{GTP} between R and C, the model allows for the possibility that receptor occupation by hormone may not be correlated in a direct linear fashion with activation of cAMP synthesis. In fact, such a linear correlation is rarely found in comparisons of agonist binding with cyclase activation (Maguire *et al.,* 1977). In several systems, maximal activation of adenylate cyclase is achieved by concentrations of agonist that lead to occupation of a small fraction of available receptors. This is exactly the expected result, if the function of hormone (and HR complex) is to initiate the activation of an individual cyclase molecule (by promoting formation of the $N_{GTP}C$ complex), while the rate of hydrolysis of GTP determines the duration of that activation, which may persist beyond occupancy of receptor by hormone. All the cell's receptors could still be required for maximal cyclase activation (as shown by experiments with β_d variants (Johnson *et al.,* 1979), described above), if the concentration of R (and thus of HR for any concentration of hormone) limits the rate of initiation of active catalytic enzyme complexes.

7.5 CONCLUSIONS

From this survey of experiments with the S49 system we can derive several useful generalizations regarding the advantages and limitations of the genetic approach for investigating hormone action. The genetic approach can provide extremely precise and elegant answers to certain questions. Thus, identification of the mutant gene product in clones resistant to cAMP or glucocorticoids demonstrated conclusively that cAMP-dependent protein kinase and glucocorticoid receptors are essential mediators of certain actions of cAMP and glucocorticoid hormones, respectively, in S49 cells.

In addition, it is clear that success of the genetic approach depends directly on the nature and degree of selective pressure than can be brought to bear on a particular cell process to produce interesting mutants. In this regard the choice of S49 cells was fortunate. The susceptibility of these cells to killing by cAMP and glucocorticoid hormones formed the basis of simple and highly effective procedures for isolating clones with mutant phenotypes.

The heuristic usefulness of genetic analysis also depends upon the extent of current knowledge, derived from biochemical and other experiments, about the molecular basis of a cellular function, such as response to a hormone. For example, the steroid-resistant S49 mutants strongly confirm but do not extend the currently prevailing model of the action of glucocorticoid receptors, including the notion that transfer of hormone—receptor complexes to the nucleus and binding to nuclear chromatin are essential steps in steroid action. These mutants do not substantially elucidate aspects of glucocorticoid action that are currently least well understood: the number, nature, and specificity of sites at which hormone—receptor complexes interact with chromatin, and the molecular events that follow this interaction. This is because the steroid-resistant mutants so far described all appear to bear lesions that affect the glucocorticoid receptor itself (see Section 7.2).

Similarly, the majority of cAMP-resistant S49 clones bear lesions affecting cAMP-dependent kinase. Their phenotypes strongly confirm the hypothesis that this enzyme mediates many (and perhaps all) effects of cAMP in mammalian cells, including regulation of growth and induction of enzymes. So far, however, the cAMP-D⁻ phenotype (Lemaire and Coffino, 1977) represents the only variant with a lesion distal to the kinase in the biochemical pathway(s) of cAMP action. This may reflect the fact that the kinase mutations are dominant in crosses with wild-type, whereas lesions affecting kinase substrates may require loss of the functions of two allelic gene products to produce cAMP resistance. If so, ingenious multi-step selection procedures may be needed to isolate mutants with lesions specifically affecting important and mysterious processes such as cAMP's regulation of cell proliferation. In the meantime, the kinase mutants provide useful tools for investigation of regulation of kinase activity (Steinberg *et al.*, 1978), the molecular interactions of kinase subunits (Hochman *et al.*, 1977), and the substrates of cAMP-dependent kinase in intact cells (Steinberg and Coffino, 1979).

Scientists habitually describe their work in terms of carefully formulated and rigorously tested hypotheses, and are reluctant to ascribe either success or failure to luck. However, it seems clear that chance played a part in determining the kinds of mutations that lead to readily isolated S49 clones resistant to glucocorticoids and cAMP, and therefore limited the relative novelty of the inferences that could be drawn from analyzing their phenotypes. The determining role of luck would be considerably mitigated if infinite resources allowed isolation and characterization of all possible classes of mutants, or if we completely understood beforehand both the biochemical systems we are investigating and their genetic control. The first possibility seems unlikely, and the second would render a genetic (or any other) investigation trivial and unnecessary. For the present at least we are stuck with the fact that luck may determine whether a particular mutant will teach us something new.

Certainly luck played an important role in determining the usefulness of S49 variants resistant to hormones that stimulate cAMP synthesis. Firstly, the frequency of appearance of *cyc⁻* variants is much higher (for unknown reasons) than that of kinase mutants; otherwise, kinase mutants might have made isolation of *cyc⁻* mutants much more difficult. Secondly, the *cyc⁻* variants did not bear a lesion that was easy to explain on the basis of current understanding of hormone-sensitive adenylate cyclase. Hormone receptors were present in normal numbers, and cAMP synthesis was unresponsive to a wide spectrum of agents that stimulate adenylate cyclase in normal cells. This led to the erroneous suggestion that the *cyc⁻* variants were deficient in catalytic adenylate cyclase activity. Correction of this error, by the ingenious reconstitution experiments of Ross and Gilman (Ross *et al.,* 1977a and b, and 1978), followed by identification of peptides labelled by cholera toxin and ³² P–NAD⁺ in wild-type but not in *cyc⁻* (Johnson *et al.,* 1978b), contributed to a real advance in our understanding of the coupling of hormone receptors to adenylate cyclase. Classical biochemical techniques would certainly have led to eventual identification of a guanyl nucleotide regulatory component of the cyclase system. In this case, however, analysis of a variant phenotype shed new light on a difficult and refractory problem.

If seems fair to conclude that somatic genetics — with a generous admixture of biochemistry, and a little bit of luck — will continue to prove useful in illuminating the mysteries of hormone action in animal cells.

REFERENCES

Abrams, A. and Smith, J. (1974), *Enzymes,* **9**, 395–420.

Bennet, V., O'Keefe, E. and Cuatrecasas, P. (1975), *Proc. natn. Acad. Sci. U.S.A.,* **72**, 33–37.

Bitensky, M.W., Wheeler, M.A., Mehta, H. and Miki, N. (1975), *Proc. natn. Acad. Sci. U.S.A.,* **72**, 2572–2576.

Blume, A.J. (1978), *Proc. natn. Acad. Sci. U.S.A.,* **75**, 1713–1717.

Bourgeois, S. and Newby, R.F. (1977), *Cell,* **11**, 423–430.

Bourgeois, S., Newby, R.F. and Huet, M. (1978), *Cancer Res.,* **38**, 4279–4284.

Bourne, H.R., Coffino, P. and Tomkins, G.M. (1975a), *J. cell Physiol.,* **85**, 611–620.

Bourne, H.R., Coffino, P. and Tomkins, G.M. (1975b), *Science,* **187**, 750–752.

Brunton, L.L., Maguire, M.E., Anderson, H.J. and Gilman, A.G. (1977), *J. biol. Chem.,* **252**, 1293–1302.

Cassel, D. and Pfeuffer, T. (1978), *Proc. natn. Acad. Sci. U.S.A.,* **75**, 2669–2673.

Cassel, D. and Selinger, Z. (1976), *Biochim. biophys. Acta,* **452**, 538–551.

Cassel, D. and Selinger, Z. (1977), *Proc. natn. Acad. Sci. U.S.A.,* **74**, 3307–3311.

Cassel, D. and Selinger, Z. (1978), *Proc. natn. Acad. Sci. U.S.A.,* **75**, 4155–4159.

Childers, S.R. and Snyder, S.H. (1978), *J. biol. Chem.,* **23**, 759–762.

Coffino, P., Bourne, H.R. and Tomkins, G.M. (1975), *J. cell Physiol.,* **85**, 603–610.

Coffino, P. and Gray, J.W. (1978), *Cancer Res.,* **38**, 4285–4288.

Friedrich, U. and Coffino, P. (1977), *Proc. natn. Acad. Sci. U.S.A.,* **74**, 679–683.

Gill, D.M. (1975), *Proc. natn. Acad. Sci. U.S.A.,* **72**, 2064–2068.

Gill, D.M. (1977), In: *Advances in Cyclic Nucleotide Research,* (Greengard, P. and Robison, G.A., eds.), 85–118.

Gill, D.M. and King, C.A. (1975), *J. biol. Chem.,* **250**, 6424–6432.

Gill, D.M. and Meren, R. (1978), *Proc. natn. Acad. Sci. U.S.A.,* **75**, 3050–3054.

Glossman, H., Baukai, A. and Catt, K.J. (1974), *J. biol. Chem.,* **249**, 664–666.

Goldin, S.M. (1977), *J. biol. Chem.,* **252**, 5630–5642.

Goldin, S.M. and Tong, S.W. (1974), *J. biol. Chem.,* **249**, 5907–5015.

Gutmann, N.S., Rae, R.A. and Schimmer, B.P. (1978), *J. cell. Physiol.,* **97**, 451–460.

Haga, T., Haga, K. and Gilman, A.G. (1977a), *J. biol. Chem.,* **252**, 5776–5782.

Haga, T., Ross, E.M., Anderson, H.J. and Gilman, A.G. (1977b), *Proc. natn. Acad. Sci. U.S.A.,* **74**, 2016–2020.

Hochman, J., Bourne, H.R., Coffino, P., Insel, P.A., Krasny, L. and Melmon, K.L. (1977), *Proc. natn. Acad. Sci. U.S.A.,* **74**, 1167–1171.

Honjo, T., Nishizaka, Y. and Hayaisha, O. (1969), *Cold Spring Harbor Symp. quant. Biol.,* **34**, 603–608.

Hyman, R. (1973), *J. natn. Cancer Inst.,* **50** (2), 415–422.

Insel, P.A., Bourne, H.R., Coffino, P. and Tomkins, G.M. (1975), *Science,* **190**, 896–897.

Insel, P.A. and Fenno, J. (1978), *Proc. natn. Acad. Sci. U.S.A.,* **75**, 862–865.

Insel, P.A., Maguire, M.E., Gilman, A.G., Bourne, H.R., Coffino, P. and Melmon, K.L. (1976), *Mol. Pharmacol.,* **12**, 1062–1069.

Ivarie, I. and O'Farrell, P.A. (1978), *Cell,* **13**, 41–55.

Johnson, G.L. and Bourne, H.R. (1977), *Biochem. biophys. Res. Comm.,* **78**, 792–798.

Johnson, G.L., Bourne, H.R., Gleason, M.K., Coffino, P., Insel, P.A. and Melmon, K.L. (1979), *Mol. Pharmacol.,* **15**, 16–27.

Johnson, G.L., Kaslow, H.R. and Bourne, H.R. (1978a), *Proc. natn. Acad. Sci. U.S.A.,* **75**, 3113–3117.

Johnson, G.L., Kaslow, H.R. and Bourne, H.R. (1978b), *J. biol. Chem.,* **253**, 7120–7123.

Kaslow, H.R., Farfel, Z., Johnson, G.L. and Bourne, H.R. (1979), *Mol. Pharmacol.,* in press.

Kyte, J. (1971), *J. biol. Chem.,* **246**, 4157–4165.

Lad, P.M., Welton, A.F. and Rodbell, M. (1977), *J. biol. Chem.,* **252**, 5942–5946.

Lemaire, I. and Coffino, P. (1977), *Cell,* **11**, 149–155.

Limbird, L.E. and Lefkowitz, R.J. (1977), *J. biol. Chem.,* **252**, 779–802.

Lin, M.C., Nicosia, S., Lad, P.M. and Rodbell, M. (1977), *J. biol. Chem.,* **252**, 2790–2792.

Littlefield, J.W. (1964), *Science,* **145**, 709–712.

Londos, C., Salomon, Y., Lin, M.C., Harwood, J.P., Schramm, M., Wolfe, J. and Rodbell, M. (1974), *Proc. natn. Acad. Sci. U.S.A., 71*, 3081–3090.

Luria, S.E. and Delbruck, M. (1943), *Genetics, 28*, 491–511.

Maguire, M.E., Van Arsdale, P.M. and Gilman, A.G. (1976), *Mol. Pharmacol., 12*, 335–339.

Maguire, M.E., Ross, E.M. and Gilman, A.G. (1977), In: *Advances in Cyclic Nucleotide Research, 8*, (Greengard, P. and Robison, G.A., eds.), pp. 1–84. Raven Press, New York.

Maguire, M.E., Wiklund, R.A., Anderson, H.J. and Gilman, A.G. (1976), *J. biol. Chem., 251*, 1221–1231.

leMaire, M., Jorgensen, K.E., Roigaard-Peterson, H. and Moller, J.V. (1976a), *Biochemistry, 15*, 5805–5812.

leMaire, M., Lind, K.E., Jorgensen, K.E., Roigaard, H. and Moller, J.V. (1978), *J. biol. Chem., 253*, 7051–7060.

leMaire, M., Moller, J.V. and Tanford, C. (1976b), *Biochemistry, 15*, 2336–2342.

Moss, J., Manganiello, V. and Vaughn, M. (1976), *Proc. natn. Acad. Sci. U.S.A., 73*, 4424–4427.

Moss, J. and Vaughn, M. (1977), *J. biol. Chem., 252*, 2455–2457.

Naya-Vigne, J., Johnson, G.L., Bourne, H.R. and Coffino, P. (1978), *Nature, 272*, 720–722.

Nimmo, H.G. and Cohen, P. (1977), In: *Advances in Cyclic Nucleotide Research, 8*, (Greengard, P. and Robison, G.A., eds.), pp. 145–266, Raven Press, New York.

Orly, J. and Schramm, M. (1976), *Proc. natn. Acad. Sci. U.S.A., 73*, 4410–4414.

Pastan, L. and Willingham, M. (1978), *Nature, 274*, 645–648.

Racker, E. (1970), In: *Membranes of Mitochondria* and *Chloroplasts* (Racker, E. ed.), pp. 135–157, Van Nostrand–Reinhold, New York.

Raeburn, S., Goor, R.S., Schneider, J.A. and Maxwell, E.S. (1968), *Proc. natn. Acad. Sci. U.S.A., 61*, 1428–1434.

Rodbell, M., Birnbaumer, L., Rohl, S.L. and Kraus, H.M.J. (1971a), *J. biol. Chem., 246*, 1977–1992.

Rodbell, M., Kraus, H.M.J., Pohl, S.L. and Birnbaumer, L. (1971b), *J. biol. Chem., 246*, 1977–1992.

Rodbell, M., Lin, M.C., Salomon, Y., Londos, C., Harwood, J.P., Martin, B.R., Rendell, M. and Berman, M. (1975), In: *Advances in Cyclic Nucleotide Research, 5*, (Greengard, P. and Robison, G.A., eds.), pp. 3–29, Raven Press, New York.

Rosen, N., Piscitello, J., Schneck, J., Muschel, R.J., Bloom, B.R. and Rosen, O. (1979), *J. cell Physiol., 98*, 125–136.

Rosenau, W., Baxter, J.D., Rousseau, G.G. and Tomkins, G.M. (1972), *Nature new Biol., 237* (70), 20–24.

Ross, E.M. and Gilman, A.G. (1977a), *Proc. natn. Acad. Sci. U.S.A., 74*, 3715–3719.

Ross, E.M. and Gilman, A.G. (1977b), *J. biol. Chem., 252*, 6966–6969.

Ross, E.M., Howlett, A.C., Ferguson, K.M. and Gilman, A.G. (1978), *J. biol. Chem., 253*, 6401–6412.

Ross, E.M., Maguire, M.E., Sturgill, T.W., Biltonen, R.L. and Gilman, A.G. (1977), *J. biol. Chem., 252*, 5761–5775.

Schramm, M. and Rodbell, M. (1975), *J. biol. Chem.*, **250**, 2232–2237.

Schramm, M., Orly, S. and Korner, M. (1977), *Nature*, **268**, 310–313.

Schwarzmeier, J.D. and Gilman, A.G. (1977), *J. cyclic nucleotide Res.*, **3**, 227–238.

Senior, A.E. (1973), *Biochim. biophys. Acta*, **301**, 249–277.

Sibley, C.H. and Tomkins, G.M. (1974a), *Cell*, **2**, 213–220.

Sibley, C.H. and Tomkins, G.M. (1974b), *Cell*, **2**, 221–227.

Simantov, R. and Sachs, L. (1975), *J. biol. Chem.*, **250**, 3236–3242.

Singer, S.J. and Nicolson, G.L. (1972), *Science*, **175**, 720–731.

Skogerson, K. and Moldave, K. (1968), *J. biol. Chem.*, **243**, 5354–5360.

Sone, N., Yoshida, M., Hirata, H. and Kagawa, Y. (1978), *Proc. natn. Acad. Sci. U.S.A.*, **75**, 4219–4223.

Steinberg, R.A. and Coffino, P. (1979), *Cell*, **18**, 718–733.

Steinberg, R.A., van Daalen Wetters, T. and Coffino, P. (1978), *Cell*, **15**, 1351–1361.

Steinberg, R.A., O'Farrell, P.H., Friedrich, U. and Coffino, P. (1977), *Cell*, **10**, 381–391.

Steinberg, R.A., Steinberg, M.G. and van Daalen Wetters, submitted.

U'Prichard, D.C. and Snyder, S.H. (1978), *J. biol. Chem.*, **253**, 3444–3452.

Vauquelin, G., Geynet, P., Hanoune, J. and Strosberg, D. (1977), *Proc. natn. Acad. Sci. U.S.A.*, **74**, 3710–3714.

Vogel, G. and Steinberg, R. (1976), *Biochemistry*, **15**, 208–216.

Yamamoto, K.R., Gehring, U., Stampfer, M.R. and Sibley, C.H. (1976), *Recent Prog. hormone Res.*, **32**, 3–32.

Yamamoto, K.R., Stampfer, M.R. and Tomkins, G.M. (1974), *Proc. natn. Acad. Sci. U.S.A.*, **71** (10), 3901–3905.

8 Some Perspectives on the Hormone-stimulated Adenylate Cyclase System

G. MATTHEW HEBDON*, HARRY LeVINE III*,
NAJI E. SAHYOUN*, CLAUS J. SCHMITGES*,
and PEDRO CUATRECASAS

* Authors' names are arranged in alphabetical order.

Abbreviations

ACTH, adrenocorticotropic hormone; EGF, epidermal growth factor; FSH, follicle stimulating hormone; Gpp(NH)p, guanylyl-5'-imidodiphosphate; G-protein, guanylnucleotide regulatory protein; hCG, human chorionic gonadotropin; LH, luteinizing hormone; NEM, *N*-ethylmaleimide; NGF, nerve growth factor; VIP, vasoactive intestinal peptide.

Membrane Receptors: *Methods for Purification and Characterization*
(*Receptors and Recognition,* Series B, Volume 11)
Edited by S. Jacobs and P. Cuatrecasas
Published in 1981 by Chapman and Hall, 11 New Fetter Lane, London EC4P 4EE

8.1 INTRODUCTION

A central question in biology concerns how regulation of cellular processes is achieved. A major aspect of this relates to the mechanism by which extracellular factors, e.g., growth factors, neurotransmitters, and hormones, exert their effects. These agents modify the intracellular status of the cell, affecting either a broad range of cell types (e.g. growth hormones) or a more restricted population (e.g. ACTH). A further characteristic is the time span within which the effects become apparent which may be long (e.g. NGF), short (e.g. cAMP generation) or both long and short (e.g. insulin effect on growth and glucose transport). An even more rapid effect is detected with those agents which seem to modulate ion-fluxes (e.g. acetyl-choline). The cell can thus show a very broad time-spectrum of responses. Long term action is believed to reflect a pleiotropic action at the level of the genome (e.g. transcription, translation, and cell division), whereas the short term effects seem to be specific alterations designed to meet particular environmental needs (e.g. cAMP production leading to glycogenolysis). This represents coarse and fine tuning of cellular metabolism, allowing the cell to efficiently co-ordinate adaptations to changing environmental conditions.

The extracellular factor, whatever its nature, may be regarded as coded information, waiting to be received and decoded by a suitably equipped cell. The question of how cell regulation is achieved may be reformulated in the following manner: How is the coded information (for growth, chemotaxis, membrane depolarization etc.). transmuted to a form which the cell can recognize and can respond to? It is possible to divide the process of information transmutation into 6 stages:

(1) recognition
(2) generation of a signal
(3) transmission of the signal to the effector system
(4) decoding of the signal by the effector
(5) response
(6) removal of signal and attenuation of responses, i.e. a return to the basal state.

Any attempt to elucidate the process of signal—effect coupling will require study of each of these steps of information transfer.

Recognition implies the presence of some component capable of discriminating the information from similar molecules encoding different messages, thus leading to the concept of a receptor.

If the response (Stage (5)) is activation of an enzyme or transport system, the products of which modulate some other enzyme system, it is apparent that the capacity for signal amplification or reduction will be present. This is a postulate of second messenger theories of hormone action, or more generally of cascade processes.

The location of receptors for the various signals has been the subject of much research. Steroid hormones can dissolve in the lipid matrix of cellular membranes and are presumed to diffuse to the cytoplasm where there is a soluble receptor. The current status of information transmission with steroid hormones has been well reviewed (Leavitt and Clark, 1979) and will not be considered further here.

The majority of hormones, neurotransmitters and growth factors do not seem capable of traversing the membrane unaided. In these cases there is ample evidence for the existence of cell surface receptors (Cuatrecasas, 1974). In situations where the effect of a hormone is discernible only after some time (e.g. EGF) the process of signal generation may involve internalization of the hormone—receptor complex (Goldstein *et al.*, 1979; Catt *et al.*, 1979) and possibly lysosomal processing (Fox and Das, 1979) to generate the signal. Subsequent events of signal transmission to the effector system are not clear but may involve generation of some cytosolic coupling factor.

In the case of fast-acting extracellular factors (neurotransmitters, glycopeptide hormones and catecholamines) there is more detailed evidence available. The nicotinic cholinergic receptor, for example, seems to show extremely rapid coupling between signal recognition and physiological effect, i.e. membrane depolarization, probably involving Ca^{2+} fluxes (second messenger) and monovalent cation channel modulation.

It cannot be excluded that events such as changes in membrane permeability, in transport enzymes or in ion fluxes occur or that they may indeed be of primary biological importance. However, in general, with the glycopeptide hormones and catecholamines, there is a rapid elevation of intracellular cAMP levels (second messenger) following hormone binding to the receptor. The subsequent physiological actions of cAMP are probably mediated by protein kinases. In this review we will concentrate on the vectorial generation of cAMP in response to hormonal stimulation.

8.2 THE RECONSTITUTIONAL APPROACH TO HORMONE STIMULATION OF ADENYLATE CYCLASE

In the last twenty years evidence has accumulated on, firstly, the number of components in the hormone-sensitive adenylate cyclase system and, secondly, the molecular mechanism governing the coupling process. This information has been difficult to obtain due to the multifactorial nature of the system. Several of the components have been found to be unstable, membrane-associated and present in very low concentration (probably < 1 pmol mg^{-1} membrane protein). Classical approaches to the resolution of complex systems by separation and purification of the components have been hampered by the difficulties noted above and also because significant hormonal stimulation of adenylate cyclase has not been found in the absence of an intact membrane structure.

Significant contributions to our understanding of the adenylate cyclase system have been made by approaches involving either differential inactivation or solubilization of components and reconstitution of some biological function. Three major approaches have been used:

(1) Type A. Fusion of cells (or membranes) deficient in different properties of the adenylate cyclase system (generated by genetic or chemical manipulation) to determine possible complementation (see below).
(2) Type B. Reconstitution of a biochemically deficient property by supplying factors which have been solubilized from another source (see below).
(3) Type C. Solubilization of the components of the adenylate cyclase system and the reconstitution of these partially purified proteins into artificial liposomes, or onto a cytoskeletal matrix (see below).

The three approaches have their respective advantages and disadvantages. Studies involving fusion experiments between cells (or membranes) that are deficient in a specific function of the adenylate cyclase complex have the advantage of not involving extensive disruption of membrane structure with the concurrent danger of denaturation of labile proteins. In membranes with different lesions the remaining components presumably complement each other by mixing during and after the fusion process. The major drawback of this approach is that it does not allow a *molecular* interpretation of the results because of the complexity of the membrane. It is impossible to unequivocally identify the components whose interaction is responsible for the reconstitution.

The second approach theoretically allows purification and molecular characterization of the factors supplied in the detergent extract; as with the fusion experiments the problem of identifying the membrane-bound components remains. This approach has been very helpful as it appears that the catalytic unit and the receptor are integral membrane proteins, requiring specific orientation in the membrane to interact with other components of the system. The proteins supplied by the detergent extract represent the coupling factors which seem to be peripheral proteins that act near the surface of the membrane.

The third approach is the only one designed to rigorously determine the minimal number of components in the adenylate cyclase system and to elucidate the molecular mechanism of transduction. This approach is also an ideal method for studying the effects of lipids on the adenylate cyclase system. There are, however, some major technical problems to this approach including the reintegration of the components into liposomes and the removal of the detergent used for solubilization since the receptor—adenylate cyclase coupling is sensitive to even low concentrations of detergents. The necessity of reconstructing a highly organized multifactorial system, possibly involving the cytoskeleton, peripheral membrane proteins and the lipid bilayer presents a considerable challenge.

8.2.1 Cell-fusion studies

The first result of the reconstitution experiments was to dispel the original model of the 'adenylate cyclase complex', according to which the catalytic unit, the hormone receptor and a hypothetical 'transducer' were integral, permanently associated parts of the complex. We now know that hormone receptors (and/or adenylate cyclase) are mobile in the plane of the membrane and are free to diffuse and interact with adenylate cyclase and/or other components involved in coupling the occupation of a hormone receptor by its ligand with the activation of adenylate cyclase. The most elegant results came from experiments of Schramm and coworkers (Orly and Schramm, 1976; Schramm *et al.*, 1977; Schulster *et al.*, 1978; Laburthe *et al.*, 1979; Schramm, 1979; Eimerl *et al.*, 1980). The kinetics described, however, may also be consistent with receptor–enzyme coupling by mechanisms other than diffusion of proteins within the bilayer.

Turkey erythrocytes were heated, or treated with NEM, to inactivate their adenylate cyclase (the β-catecholamine receptors were not affected by this treatment). These pretreated cells were fused with Friend erythroleukemia cells that possessed adenylate cyclase but not β-adrenergic receptors. Fusion of the two cell types resulted in functional coupling of the β-adrenergic receptor from one cell type with the adenylate cyclase from the other cell type (adenylate cyclase activity in ghosts of the fused preparation could be stimulated significantly by isoproterenol within a few minutes after fusion). Inhibitors of protein synthesis had no effect on the reconstitution of isoproterenol-stimulated adenylate cyclase activity, suggesting that coupling occurred between pre-existing components. These studies have been extended substantially by Schramm and colleagues and by other workers (Schwarzmeier and Gilman, 1977; Dufau *et al.*, 1978). It was shown that β-adrenergic, prostaglandin, glucagon, LH, and VIP receptors could be donated and coupled to heterologous adenylate cyclase systems among a number of cell types with diverse origins. These exchanges were shown between intact cells, between intact cells and isolated membranes, intact cells and receptor-containing liposomes, and between different preparations of isolated membranes.

8.2.2 Reconstitution with soluble extracts

(a) *Biochemical separation of components*
The fact that hormone receptors and the catalytic unit of adenylate cyclase are separate proteins was subsequently also shown by their physical separation by gel filtration and sucrose density centrifugation (Haga *et al.*, 1977a; Limbird and Lefkowitz, 1977).

The experiments described above clearly demonstrated that several hormone receptors and the catalytic unit of adenylate cyclase are separable entities and that a receptor derived from one cell type can functionally couple with the catalytic adenylate cyclase moiety derived from another cell type. However, the process of activation of adenylate cyclase by hormone receptors does not seem to involve a simple binding of the occupied receptor to the catalytic unit of adenylate cyclase. The activation of adenylate cyclase by hormones is markedly increased by (and in most systems has been shown to be strictly dependent upon) the addition of GTP (Rodbell *et al.*, 1971 and 1975; Ross *et al.*, 1977). Gpp(NH)p and GTPγS, GTP-analogs whose terminal phosphate is resistant to hydrolysis by cellular GTPases, activate adenylate cyclase in the absence of hormone in a quasi-irreversible manner (Londos *et al.*, 1977; Schramm and Rodbell, 1975; Pfeuffer and Helmreich, 1975). Hormones that activate adenylate cyclase accelerate the activation by Gpp(NH)p or GTPγS but do not affect the maximum activity achieved. These findings led to the hypothesis that the activation of adenylate cyclase by hormones is controlled by a guanylnucleotide regulatory component (G-protein) that is separate from the catalytic unit and the hormone receptor (Rodbell *et al.*, 1975; Maguire *et al.*, 1977).

It is presumed that hormone binding to its receptor causes an exchange of bound GDP for GTP (Cassel and Selinger, 1978); with GTP bound the G-protein can activate adenylate cyclase. A hormone-stimulated GTPase activity, first described by Cassel and Selinger, provides the mechanism to return the system to the basal state (Cassel and Selinger, 1976 and 1977; Cassel *et al.*, 1977 and 1979). Direct evidence for the existence of such a G-protein, and clues to its role in the stimulation of adenylate cyclase by hormones and cholera toxin, came first from the work of Pfeuffer and Helmreich (Pfeuffer and Helmreich, 1976; Pfeuffer, 1977). Using a GTP-Sepharose affinity matrix they were able to separate a detergent-solubilized preparation from pigeon erythrocytes containing adenylate cyclase into two fractions, a catalytic unit (which did not bind to the affinity matrix) and a G-protein which was retained by the GTP-Sepharose. The catalytic unit, further purified by covalent chromatography on disulfide-Sepharose, was not stimulated by Gpp(NH)p, NaF, or Mg^{2+}; its activity could only be assayed in the presence of Mn^{2+}. Addition of a fraction released from GTP-Sepharose with GTP (containing a 42 000 mol. wt. protein) to the catalytic unit reconstituted stimulation of adenylate cyclase activity by Gpp(NH)p, NaF, or Mg^{2+}. Reconstitution experiments with the GTP-binding fraction obtained from isoproterenol-treated membranes suggested that the hormonal activation of adenylate cyclase by guanylnucleotides is mediated via the G-protein.

Important progress towards identification of the guanylnucleotide-binding protein and the molecular mechanism whereby it regulates adenylate cyclase activity came from work with cholera toxin. Cholera toxin, an enterotoxin of *Vibrio cholerae,* irreversibly activates adenylate cyclase in mammalian tissues and broken cell preparations (Bennett and Cuatrecasas, 1976; Gill and King, 1975). Activation in broken cell preparations is dependent on ATP, GTP, cellular cytosolic factors and NAD^+ (Gill, 1977; Vaughan and Moss, 1978; LeVine and Cuatrecasas, 1980a and b). Using $[\alpha^{32}P]$-labeled NAD^+ in analogy to work with diphtheria toxin, Cassel and Pfeuffer (1978), and Gill and Meren (1978) were able to show cholera toxin-catalyzed incorporation ^{32}P label into several membrane proteins, among them a 42 000 mol. wt. protein. Extraction of cholera toxin-treated membranes with

Table 8.1 Summary of sedimentation data for solubilized adenylate cyclase components*

Ligand	Adenylate cyclase catalytic subunit	G-protein	Complex formed by catalytic subunit and G-protein
Mg^{2+}	6.0S	5.5S	No complex
Mg^{2+}/GTPγS, Gpp(NH)p	6.0S	3.4S	7.6S (7.4S)
Mg^{2+}/GDP	6.0S	5.5S	No complex
Mg^{2+}/NaF	6.0S	—	7.6S (7.4S) (preactivated)

*　Data from Pfeuffer (1979) and Guillon *et al.*, (1979).

detergent followed by affinity chromatography on a GTP-Sepharose column allowed Cassel and Pfeuffer to extend previous work and obtain significant purification of the 42 000 mol. wt. labeled protein, its complete separation from the other labeled proteins and from adenylate cyclase. The fraction containing the purified ADP-ribosylated GTP-binding moiety from cholera toxin-treated membranes conferred a characteristic, enhanced GTP-stimulated activity on adenylate cyclase solubilized from nontreated membranes. The cholera toxin-induced stimulation of adenylate cyclase activity and the incorporation of ^{32}P-label into the 42 000 mol. wt. protein were partially reversed upon incubation with cholera toxin and nicotinamide at pH 6.1. This indicates that cholera toxin activates the adenylate cyclase system by catalyzing an ADP-ribosylation of the 42 000 mol. wt. G-protein. Cassel and Selinger had previously found that cholera toxin inhibits the catecholamine-stimulated GTPase activity in turkey erythrocyte membranes (Cassel and Selinger, 1976; Cassel *et al.*, 1977). This is consistent with the finding that in cholera toxin-treated membranes GTP becomes as effective as Gpp(NH)p in stimulating the enzyme. Based on this, Pfeuffer and Cassel have suggested that the 42 000 mol. wt. G-protein itself reversibly associates with the catalytic unit of adenylate cyclase and is the GTPase activity which governs the association–dissociation cycle. However, demonstration of GTPase activity in the purified guanylnucleotide-binding protein has not yet been reported.

Using the G-protein, labelled with [^{32}P] ADP-ribose by cholera toxin and purified by GTP-Sepharose chromatography, it was possible to demonstrate that the sedimentation coefficient of the G-protein was dependent on the type of guanylnucleotide bound and that the catalytic unit and the G-protein co-sedimented only in the presence of guanylnucleotide triphosphate (Pfeuffer, 1979) (see Table 8.1). Although GDP-liganded G-protein was incapable of associating in a stable form with the adenylate cyclase, NaF prestimulation of the membrane-bound enzyme produced a rapidly sedimenting form of the enzyme. Enzyme fully stimulated by cholera toxin, whose activity is refractory to NaF, will not form the 7.6S species in the presence of Mg^{2+}/NaF, implying that the conformation of the modified G-protein is not affected by NaF

Similar observations were subsequently made with the pig kidney medulla adenylate cyclase (Guillon *et al.*, 1979). This work provided the additional information that the 6.0 and 7.4S forms of the enzyme have indistinguishable Stokes radii. It appeared that nucleotide binding altered detergent binding sufficiently to compensate for the increase in molecular weight. In these studies NaF caused no change in the calculated, detergent-free, molecular weight of the enzyme, although a change in the apparent sedimentation coefficient of the NaF-prestimulated enzyme was observed.

The comigration on gel filtration of basal, Gpp(NH)p- or NaF-stimulated, digitonin-solubilized frog erythrocyte adenylate cyclase has also been noted (Limbird *et al.*, 1979).

(b) *Mutant cells – the S49 lymphoma system*

The experiments previously described involve a chemical manipulation and dissection of the adenylate cyclase system (differential inactivation or solubilization of components or total solubilization and separation of the solubilized components). Feeling that *genetic* depletion of the adenylate cyclase system would be more specific and could result in the total inactivation of a component, the groups of Gilman *et al.* and Bourne *et al.* have used an approach based on the concepts and techniques of somatic genetics. They made use of the original observation of Tomkins that S49 murine lymphoma cells are killed by high intracellular cAMP levels, and his technique of selection for mutants defective in the hormone receptor–adenylate cyclase–protein kinase system. Selection of S49 cells under conditions that lead to a stimulation of endogenous cAMP production (presence of hormones or cholera toxin) has allowed the isolation of four classes of mutants of the hormone receptor–adenylate cyclase system:

(1) AC^- (or cyc^-) (Bourne *et al.*, 1975). The adenylate cyclase in membranes derived from this strain cannot be stimulated by hormones, Gpp(NH)p, NaF or Mg^{2+}. Basal adenylate cyclase activity can be assayed, however, in the presence of Mn^{2+} (Ross *et al.*, 1978; Naya-Vigne *et al.*, 1978). AC^- cells contain the same number of β-adrenergic receptors as the wild type (Insel *et al.*, 1976). All indications are that the G-protein is defective or absent in AC^- (Ross and Gilman, 1977a and b; Johnson *et al.*, 1978a and b).

An S49 variant truly deficient in the catalytic unit of adenylate cyclase has yet to be isolated. However, a hepatoma cell line, HC–1, has been found which contains no detectable adenylate cyclase activity even when assayed in the presence of Mn^{2+} while retaining the 42 000 mol. wt. G-protein and the β-adrenergic receptor (Ross *et al.*, 1978).

(2) *UNC* (Haga *et al.*, 1977b). In membranes derived from this strain hormones (catecholamines, PGE_1) do not stimulate adenylate cyclase activity. The adenylate cyclase activity is, however, stimulated by Gpp(NH)p, NaF and Mg^{2+}. UNC cells contain the same number of β-adrenergic receptors as the wild type (or even slightly more),

so the defect in UNC seems to involve an adenylate cyclase—hormone receptor coupling factor, but the exact locus (or loci) of the mutation is not known. The AC^- and UNC mutations do not complement each other (Schwarzmeier and Gilman, 1977; Naya-Vigne *et al.*, 1978). That could mean that both AC^- and UNC lack a component that is essential for coupling of hormone receptors to adenylate cyclase, and that the AC^- variant in addition lacks another one that is necessary for Gpp(NH)p-, NaF- and Mg^{2+}-stimulation. Alternatively, both mutations could modify to different degrees a component necessary for both coupling and Gpp(NH)p-, NaF- and Mg^{2+}-stimulation of adenylate cyclase. According to this hypothesis the AC^- mutation causes the loss of both functions while the UNC mutation would only affect the coupling between hormone receptors and adenylate cyclase.

(3) Rec^d (Johnson *et al.*, 1979). These cells contain a drastically reduced number of β-adrenergic receptors and show no (or only slight) stimulation of adenylate cyclase by catecholamines. Stimulation of the enzyme by PGE_1, Gpp(NH)p, NaF and Mg^{2+} is not changed compared with the wild type. S49 variants of this type have been isolated only recently. Therefore, in the earlier reconstitution experiments a murine L cell line, B82, with PGE_1-, Gpp(NH)p-, NaF- and Mg^{2+}-sensitive adenylate cyclase activity but no β-adrenergic receptors (by the criteria of binding or stimulation of adenylate cyclase) was used (Ross and Gilman, 1977a).

(4) *Various protein kinase mutants* (Daniel *et al.*, 1973; Insel *et al.*, 1975). These should be useful for further studies on the biological effects of cAMP and to study the role of protein phosphorylation.

The variants described above have been used for reconstitution experiments primarily of the Types A and B (fusion and addition of detergent extracts to membranes) (Schwarzmeier and Gilman, 1977; Ross *et al.*, 1978; Naya-Vigne *et al.*, 1978; Insel *et al.*, 1976; Ross and Gilman, 1977a and b; Johnson *et al.*, 1978; Kaslow *et al.*, 1979, Howlett *et al.*, 1979; Sternweis and Gilman, 1979). The fusion experiment showed that AC^- and UNC are recessive to the respective wild type alleles and that AC^- and rec^d and UNC and rec^d complement each other while UNC and AC^- do not complement. Addition of detergent extracts to membranes led to exciting but ambiguous results. Recombination of a hormonally sensitive adenylate cyclase system has proven possible only when the membranes contain the catalytic unit and the hormone receptor. Hormonal responsiveness could not be achieved in the soluble state. Addition of solubilized adenylate cyclase catalytic unit to a membrane containing receptor and coupling factors, or addition of solubilized receptor to a membrane containing the catalytic unit and coupling factors, did not restore hormonally sensitive adenylate cyclase activity. In contrast, incubation of membranes derived from AC^- or UNC cells (i.e., membranes containing catalytic unit and receptors) with detergent extracts from either wild type, HC–1 or rec^d cells (i.e. containing coupling factors) leads to a reconstitution of adenylate cyclase activity that can be stimulated by hormones, Gpp(NH)p, NaF and Mg^{2+}. The most probable explanation is that the catalytic unit and receptors are integral membrane proteins (they can be solubilized only by detergents and when solubilized bind a large amount of detergent). They must also be oriented in the membrane in a specific fashion to interact productively with one another and with other components of the hormone receptor–adenylate cyclase system. In contrast, the coupling factors behave more as peripheral proteins (they can be solubilized by urea and do not bind detergent) that can interact with the catalytic unit and hormone receptors at the surface of the membranes. The mode of interaction of these components with the plasma membrane and cytoskeleton will be considered in a later section.

Markedly different results were obtained when different detergents were used for the solubilization of the coupling factors. The simple mixing of Lubrol PX-solubilized coupling factors and membranes derived from AC⁻ cells did not result in reconstitution which was stable to dilution and washing of the membranes. Persistent reconstitution was only found when the detergent-solubilized coupling factors were incubated with membranes in the presence of activators of the enzyme, NaF + ATP, or Gpp(NH)p (or GTP when the coupling factors were derived from cholera toxin-treated cells). All these effectors seem to exert their action on the coupling factors; they do not seem to modify directly the catalytic unit of adenylate cyclase. This is consistent with the observation that the G-protein activity (see below) is stabilized against thermal inactivation by the presence of Gpp(NH)p (Ross and Gilman, 1976b). During stable reconstitution coupling factor activity disappears from the supernatant and this removal is concomittant with the appearance of a stimulated adenylate cyclase activity in the membrane. Coupling factor activity can be extracted by detergent from reconstituted AC⁻ membranes in a form that is indistinguishable (by sucrose density gradient centrifugation) from the original soluble coupling factor activity.

Cholate-solubilized coupling factors which retain no measurable adenylate cyclase activity can be incorporated into AC⁻ or UNC membranes even in the absence of Gpp(NH)p or NaF (Sternweis and Gilman, 1979). This technique facilitates the study of reversible activation of the reconstituted adenylate cyclase system by hormones. AC⁻ membranes reconstituted with cholate-solubilized coupling factors display hormonal response to PGE_1 and catecholamines similar to those of wild type S49 cell membranes. Wild type and reconstituted membranes show a similar dependence on agonist concentrations for both adenylate cyclase activation and receptor binding. In UNC and AC⁻ membranes β-agonist binding is insensitive to guanylnucleotides, whereas in reconstituted as in wild type membranes the affinity of catecholamines for the β-receptor is decreased by guanylnucleotides.

The reasons for the markedly different results obtained with the two detergents are not clear. One possibility is that Lubrol has a higher affinity than cholate for factors necessary for reconstitution and may therefore interfere with the interaction of the component of the hormone receptor–adenylate cyclase system. Lubrol may also labilize the membrane structure because of its tendency to partition into the bilayer; this labilization may impair stable incorporation of the factors into the membranes.

8.2.3 Reconstitution studies with partially purified components

In this laboratory the interactions of various modulatory factors with adenylate cyclase have been studied. The following discussion presents the progress made in this area.

Bradham (1977) had published a report on a protein factor which appeared to mediate NaF-stimulation of adenylate cyclase and could be solubilized from rat brain with detergent. Readdition of this soluble factor reconstituted a NaF-responsive enzyme. In our studies (Sahyoun *et al.*, 1977; Hebdon *et al.*, 1978) we have also used rat brain, which has high adenylate cyclase activity and is available in reasonably large quantity. A Gpp(NH)p- and NaF-insensitive enzyme was produced by sequential glycerol, sucrose and detergent extractions. This enzyme could be stimulated by Gpp(NH)p or NaF only following incubation with detergent extracts of the particulate fractions from a variety of tissues. The tissue source we have used principally is adipocytes from rat epididymal fat pads. A comparably Gpp(NH)p- and NaF-insensitive adenylate cyclase has also been chromatographically prepared and subsequently reconstituted (Drummond *et al.*, 1980).

The non-responsive enzyme from rat brain was similarly activated by regulatory components in either the particulate or soluble state. Activation did not occur at $0°$ and required the simultaneous presence of the following: adenylate cyclase, one (or more) solubilized regulatory proteins, $MgCl_2$ and ligand (Gpp(NH)p or NaF). Despite permanent activation in the presence of Gpp(NH)p or NaF, no direct evidence for ligand-induced binding of a solubilized protein factor to the enzyme has been obtained. However, strong indirect evidence is available that activation of the particulate enzyme by soluble factors and ligand can be reversed by detergent extraction. Restimulation of this preparation required the re-addition of regulatory proteins.

Since the initial report a number of treatments have been shown to selectively destroy the Gpp(NH)p and NaF responses of membrane-bound adenylate cyclase. Exposure of rat cerebellar synaptosomes to millimolar concentrations of formaldehyde or glutaraldehyde abolished guanylnucleotide and NaF stimulation; at higher concentrations EGTA inhibition of the enzyme was prevented and the basal activity was elevated (Monneron and d'Alayer, 1980). In contrast, inhibition of the basal enzyme activity by Ca^{2+} or adenosine and its analogues was unaltered. Reconstitution of this system using exogenous coupling factors was not reported.

An interesting report has appeared showing that reconstitution of guanylnucleotide- and NaF-sensitive adenylate cyclase is not restricted to mammals and birds. A detergent-solubilized preparation containing heat- or NEM-inactivated adenylate cyclase from *Fasciola hepatica* (the liver fluke) can reconstitute a partial NaF and GTPγS response with solubilized, active enzyme (Renart *et al.*, 1979).

(a) *Properties of the regulatory components*
Solubilized regulatory components were trypsin-sensitive, heat-labile and were inactivated by 1 mM NEM or by brief exposure to pH below 5. In contrast,

membrane-bound factors were stable to treatment with 100 μg ml^{-1} trypsin for 10 minutes at 30°C. Membrane-bound factors were also stable to high salt (0.5–1.0 KCl), 0.1 mM EDTA, and 1M urea, *C. welchii* phospholipase C (2 units ml^{-1} for 10 min at 30°C).

The regulatory components could not be solubilized in an activated state from Gpp(NH)p- or NaF-pretreated membranes. Isoproterenol (50 μM), glucagon (10 μM) and insulin (0.1–1 mU ml^{-1}) had no effect on the activity of the regulatory components either added directly to the reconstitution systems or to the fat cell membranes prior to solubilization. Reconstitution of Gpp(NH)p and NaF responses was similarly unaffected by cytochalasin B (1 μM), colchicine (100 μM), calcium-dependent regulator protein (100 μg ml^{-1} ± 100 μM Ca^{2+}) or platelet smooth muscle actin or actinomyosin.

(b) *Mechanism of interaction of regulatory components with adenylate cyclase*
The relatively simple reconstitution procedure described above allows a partial clarification of the molecular interactions occurring. A similar interpretation was subsequently made for the S49 lymphoma system. The regulatory protein(s) of fat cell membranes clearly contain distinguishable NaF- and Gpp(NH)p-reconstituting activities as shown by their differential trypsin sensitivities and lack of cross-stabilization against thermal inactivation. Since we have been unable to clearly separate the two activities by physico-chemical techniques we cannot distinguish between the following alternatives:

(a) the activities are simultaneously present on one molecule, or
(b) the activities represent different, but exclusive, conformational states (conformers) of the same molecule, or
(c) the activities are present on two separate and distinct molecules which have not yet been resolved.

It seems clear that the activating ligands can interact with their respective regulatory components, as can be judged by their increased trypsin sensitivity and decreased thermal lability in the presence of ligand. The ligand-induced protection against thermal inactivation can occur in the absence of divalent cations. This interaction of ligand with the regulatory component is rapidly reversible as there is no persistent activation after removal of Gpp(NH)p or NaF.

The reconstituted adenylate cyclase system shows persistent activation, a time course and concentration dependence of activation comparable with the unextracted membrane-bound enzyme, a requirement for MgCl$_2$ and temperatures above 0°C. As the activation can be reversed by detergent extraction and again reconstituted by the re-addition of exogenous protein factors, it seems unlikely that the catalytic component is covalently modified during activation. The separation of catalytic from regulatory components discussed here may include some thermal inactivation of the latter. This would imply that the regulatory components were physically *in situ* but inactive. However, the demonstration of re-activation implies that any inactive endogenous factor must be readily displaceable by active, exogenous protein. The kinetics of activation are also consistent with a non-enzymatic mechanism for the stimulation of adenylate cyclase by Gpp(NH)p or NaF.

The observations above are consistent with the following model for the activation of adenylate cyclase:

$$P_R + L \overset{fast}{\to} L \cdot P_R \qquad [1]$$

$$L \cdot P_R + E + MgCl_2 \overset{slow}{\to} L \cdot P_R \cdot E^* \qquad [2]$$

in which P_R represents the regulatory component(s), E the catalytic units (E^* is the activated state) and L the ligand (either Gpp(NH)p or NaF). Hormones increase the rate at which Gpp(NH)p activates the enzyme (Cuatrecasas *et al.,* 1975; Jacobs and Cuatrecasas, 1976), presumably by promoting the dissociation of GDP from the guanylnucleotide regulatory site (Cassel and Selinger, 1978); that is, hormones act at step [2] in the above scheme. Step [2] is probably a high affinity, low capacity reaction since most of the regulatory components are recoverable free in solution after full activation of the brain adenylate cyclase. This further suggests that the factors are present in great excess over the catalytic component. The significance of this nonstoichiometric relationship between catalytic unit and regulatory proteins is not known but may suggest that the regulatory factors play other roles in the cell in addition to stimulating adenylate cyclase. Presumably one, or more, of the regulatory proteins is related to the hormone-sensitive GTPase demonstrated in avian erythrocytes (Cassel and Selinger, 1977), although this remains to be confirmed experimentally. The above model is based on data obtained with a relatively crude system and will be refined as more detailed information becomes available.

An additional putative function for the guanylnucleotide binding protein is in the mediation of calmodulin-dependent Ca^{2+} stimulation of brain adenylate cyclase (Toscano *et al.,* 1979). The enzyme could be made insensitive to Gpp(NH)p, NaF and calmodulin $\cdot$ Ca^{2+} by passing detergent-solubilized material over a calmodulin-Sepharose column. The non-responsive enzyme had no affinity for this matrix, even in the presence of Ca^{2+}. Reconstitution of the response to activators was obtained by incubating the non-responsive enzyme with a detergent extract of brain membranes, divalent cation and ligand. The major conclusion from this work is that guanyl-nucleotide was required to observe reconstitution of the calmodulin $\cdot$ Ca^{2+} response thus implicating the G-protein in the effects of calmodulin on adenylate cyclase. Further work showing either the unequivocal restriction of the calmodulin $\cdot$ Ca^{2+} response to neural tissues or the transfer of this activity to normally non-sensitive tissues will be important in evaluating the broader implications of these observations.

(c) *Reconstitution of solubilized receptors*
The receptor portion of the hormone-specific membrane signaling system has proven difficult to work with. Although hormone binding *per se* was relatively tolerant to solubilization by certain detergents, and solubilized receptor could be assayed by

the ligand binding reaction, the coupling between receptor and adenylate cyclase activity was markedly sensitive to detergent and other perturbations, with hormonal effects disappearing before solubilization of binding activity was achieved. Reconstitution of the hormonal responsiveness of adenylate cyclase has been achieved in a small number of cases. PGE_1 and catecholamine response have been reconstituted in AC^- membranes which contain pre-existent hormone receptors as discussed in another section.

There are at present three examples of systems where the receptors is first removed from the membrane and then reinserted into an appropriate target membrane.

(i) *Gonadotropin receptor.* Dufau *et al.* (1978) isolated adenylate cyclase-deficient triglyceride vesicles containing 5–10% of the luteinizing hormone (LH) receptor sites of a rat ovary preparation. When these vesicles were incubated with dissociated rat adrenal fasciculata cells a progressive incorporation of LH binding sites into the adrenal cells was noted. These cells, which originally did not have specific LH binding sites and did not respond to gonadotropic hormones, responded to 1 nM hCG with a stimulation of cAMP production coupled to increased steroidogenesis. Although the carrier lipid vesicles appeared in the cytosol of the cells, all of the LH receptors remained accessible to ^{125}I-hCG at the external surface of the cell. An evaluation of the minimum number of components necessary for hormonal coupling is not possible with these data as a number of coupling factors may have been present in the crude vesicle fraction.

(ii) *β-adrenergic receptors.* A second example is the transfer of a functional catecholamine receptor to an adenylate cyclase-containing system (Eimerl *et al.*, 1980). β-Adrenergic receptor was solubilized from turkey erythrocyte plasma membrane using deoxycholate, phospholipid was added and the detergent removed by pelleting the receptor and lipid. This receptor was subsequently implanted into the membranes of Friend erythroleukemia cells (FEC) which respond to PGE_1 but do not have any measurable β-receptor binding or response. Following implantation the adenylate cyclase of FEC was stimulated 30-fold by isoproterenol, compared with an activation by PGE_1 of 15-fold. Stimulation by PGE_1 or NaF was not significantly affected by the fusion process. The implanted receptor showed the stereospecificity and inhibitability by propranolol characteristic of β-receptor. Although the receptor was not shown to be associated with the lipid vesicles before fusion (e.g., by equilibrium density measurements), the implantation of the receptor into FEC membranes required the presence of phospholipids in the fusion mixture.

This is the only example of a reconstitution where the receptor has been solubilized, in an attempt to reduce possible contamination with other components, before combination with a separate source of adenylate cyclase and coupling factors. It should be pointed out, however, that the $[^{32}P]$-ADP-ribosylated G-protein is clearly solubilized along with a large number of other proteins from pigeon erythrocyte membranes under conditions similar to those described in this work (unpublished results), and so the G-protein may have been present in the β-receptor preparation. Further characterization of the 'liposomal' β-receptor and extension of the technique to include fusion of adenylate cyclase containing- and β-receptor-containing liposomes may allow a more biochemical description of hormonal stimulation of adenylate cyclase.

(iii) *Dopamine receptors.* Two laboratories have independently reported the successful reconstitution of dopamine (D–1)-sensitive adenylate cyclase from the caudate nucleus. Sano *et al.* (1978, 1979) first reported the reconstitution of a soluble dopamine-sensitive adenylate cyclase from canine caudate nucleus. Lubrol PX was used to solubilize enzyme and receptor from caudate synaptosomes; NaF was present during solubilization to stabilize the catalytic unit. The detergent extract was centrifuged at 105 000 g and the supernatant applied to a Sephadex G-200 column, to simultaneously remove detergent and NaF, and to resolve adenylate cyclase activity from the peak of [^{3}H] dopamine binding. Recombination at 4°C of the adenylate cyclase unresponsive to dopamine but stimulable by NaF and Gpp(NH)p, with the [^{3}H] dopamine binding fraction (K_a 3.2 μM) revealed a concentration- and GTP-dependent stimulation of adenylate cyclase activity (K_{act} 1.6 μM). Stimulation was inhibited by the neuroleptic drugs, chlorpromazine (K_i 0.62 μM) and haloperidol (K_i 0.06 μM), but not by β-agonists, -antagonists or by a monoamine oxidase inhibitor. Addition of lipid was not required for reconstitution, although the presence of small amounts of endogenous lipid in the preparation was not rigorously excluded.

Interestingly, the adenylate cyclase-containing fraction from gel filtration contained 10.5 pmol mg^{-1} protein of Gpp(NH)p-binding activity while the [^{3}H] dopamine binding peak contained none. Although a fortuitous comigration of catalytic activity and regulatory proteins cannot be excluded, the observation may have some relevance to the question of separate catalytic unit- and receptor-linked G-proteins.

Another laboratory has also reported the reconstitution of dopamine-sensitive adenylate cyclase (Hoffman, 1979a and b). The receptor and catalytic unit solubilized in cholate were separated either by gel filtration or DEAE-cellulose chromatography and could be reconstituted, before or after separation, by ammonium sulfate precipitation in the presence of crude soybean phospholipid. The component conferring dopamine sensitivity eluted from the DEAE-cellulose column at low salt concentrations whereas the basal and guanylnucleotide-stimulable enzyme activity eluted with 1 M KCl. The reconstituted adenylate cyclase was stimulated 3–4 fold by dopamine (K_{act} 8 μM) or norepinephrine (K_{act} 100 μM) and 8–12 fold by Gpp(NH)p. The activation by dopamine was inhibited by micromolar concentrations of fluphenazine, chlorpromazine or haloperidol, neuroleptic drugs which are classical blockers of the low affinity (D–1) adenylate cyclase-linked and the high affinity (D–2) non-adenylate cyclase-linked dopamine binding sites. The success of this procedure may be due to the presence of dopamine during solubilization and the use of ammonium sulfate to produce high protein concentrations.

The two independent reconstitution methods for dopamine-dependent stimulation of adenylate cyclase suggest that this may be a fruitful area of study. Further work involving purification of the components remains to be done, but the advances made so far in terms of solubilizing and reconstituting hormonal sensitivity independent of the membrane cannot be overemphasized. The results of further work in this area are eagerly awaited.

8.3 EVENTS SUBSEQUENT TO HORMONE BINDING TO RECEPTORS

The primary event following binding of hormone to receptor in adenylate cyclase-coupled systems is presumably a change in conformation and/or association state of the receptor molecule. Hormonal effects on the activation of adenylate cyclase may depend on the electrical properties of the plasma membrane e.g., membrane potential (Grollman *et al.*, 1977; Wagner and Davis, 1979). There are some other early effects that may or may not be coupled to this process, e.g. ion fluxes.

Although monovalent cation fluxes do not seem to be altered during the early phases of hormonal stimulation, divalent cation movements do seem to change. $^{28}Mg^{2+}$ but not $^{45}Ca^{2+}$ accumulation by S49 lymphoma cells is inhibited by β-agonists or PGE_1 stimulation (Maguire and Erdos, 1978). Propranolol was shown to reverse the inhibition induced by β-agonists. Studies with mutant lymphoma cells deficient in either the G-protein (AC^-) or in an undefined hormonal coupling component possibly distinct from the G-protein (UNC) showed that $^{28}Mg^{2+}$ accumulation in these cells was independent of added agonist. Both of the mutant cell lines had a normal affinity and binding capacity for hormone. The hormonal effects are not mediated via cAMP since treatment of cells with cholera toxin or with a membrane-permeant analog of cAMP had no effect on the $^{28}Mg^{2+}$ uptake which could be inhibited by β-adrenergic agonist or PGE_1 (Maguire and Erdos, 1980). Effects of β-agonists on ion fluxes seem to occur immediately after binding and may therefore possibly play a role in signal transmission through the membrane. Mg^{2+} but not Ca^{2+} increases the affinity of the β-receptor for agonists without changing the K_{act} of wild type cells (Williams *et al.*, 1978; Bird and Maguire, 1978); there is no effect on agonist affinity in the mutant AC^- or UNC cells. The effects of Mg^{2+} on the affinity of wild type cells for agonist are not expressed in the presence of GTP. It is possible that tightly bound guanyl nucleotide may explain why agonist binding in turkey erythrocytes does not respond to added Mg^{2+} (Lad *et al.*, 1980). Mg^{2+} may also be involved in events related to hormone–adenylate cyclase coupling of which we currently have little information.

Another possible mechanism of signal transmission would be modulation of the membrane potential as has been shown for thyroid stimulating hormone (TSH) binding to thyroid tissue (Grollman *et al.*, 1977). Such a variation in membrane potential may directly influence the activities of membrane-bound enzymes or influence the extent to which an enzyme is buried within the lipid matrix, i.e. its accessibility. The study of ionic- or membrane potential-controlled events by reconstitution techniques requires an impermeant matrix. This necessitates the use of resealed membranes or liposomes; the development of a liposomal system suitable for these investigations will be considered in a later section.

8.4 MULTIMERIC STATE OF THE ADENYLATE CYCLASE SYSTEM AS ASSESSED BY TARGET ANALYSIS

The interactions between components of the hormone-sensitive adenylate cyclase should lead to transient changes in the apparent molecular weights of the hormone receptor, catalytic unit and G-protein coincident upon their forming complexes.

The ingenious technique of target size analysis has provided information about possible associations of components of the adenylate cyclase couple in the intact membrane (Houslay *et al.*, 1977; Schlegel *et al.*, 1979; Martin *et al.*, 1979). This technique relies on irradiation by high energy (15 MeV) electrons to produce ionization-dependent destruction of protein function. The ionization events occur randomly throughout the uniformly irradiated sample with the full energy of the ionization being propagated through the covalently bonded atoms of the molecule. The functionality of the target molecule is destroyed by a 'single hit'. A key observation is that non-covalent bonding i.e. ionic or hydrophobic protein–protein interactions will not transmit the destructuve energy. Since the ionizations are random events, the probability of a 'hit' within a molecule is proportional to its average, unhydrated spherical, molecular volume (α molecular size). If a given reaction requires the participation of a series of components the probability of inactivation will be determined by the sum of the sizes of the individual components. This conclusion assumes that no exchange of active components with those inactivated by ionization occurs. After irradiation of separate frozen samples at a series of radiation doses the membranes are thawed and assayed for adenylate cyclase activity in the presence of the ligand being investigated. The residual activity as a function of the irradiation dose can be used to calculate the size of the functional unit; this represents the size of the minimum assembly of components necessary for a given activity. Preincubation of the membranes with ligand before irradiation can also provide information about ligand-induced changes in aggregation. The validity of this type of approach, for a variety of soluble and membrane bound proteins, has been verified for several defined systems (Houslay *et al.*, 1977).

The results obtained by this technique are only approximate due to technical problems and some assumptions which are made. However, the information summarized by Rodbell (1980) is provocative. According to the model, binding of an agonist to its receptor leads to a disruption of a complex equilibrium among oligomers of the G-protein, catalytic unit and hormone receptor. Rigorous testing of this concept will require the dissection and functional reconstitution of the separate components of the system.

8.5 PARTICIPATION OF CYTOSOLIC FACTORS

The results discussed so far have referred mainly to the cell membrane and the question of possible cytosolic involvement in the process arises. A number of variously defined cytosolic factors or components have been described and partially purified from the cytosol of different tissues. The mode of action of these factors seems to involve stabilization or augmentation of the basal adenylate cyclase activity and potentiation of the activation induced by hormones, cholera toxin and NaF. Cytosolic factors from rat liver (Pecker and Hanoune, 1977a and b; Katz *et al.*, 1978) and heart (Sanders *et al.*, 1977) have been described which increase both basal and hormone- (glucagon and catecholamine) stimulated activities. Pretreatment of fat cell membranes with cholera toxin seemed to affect the ability of the cytosol to augment epinephrine stimulation in fresh membranes (Ganguly and Greenough, 1975). Pecker and Hanoune (1977a), described a factor from red cell cytosol that restored epinephrine and NaF responsiveness to rat reticulocyte membrane adenylate cyclase. This is in accord with similar observations made in this laboratory on the same system. Equally intriguing results were obtained by treating lymphocyte membranes with immobilized Concanavalin A to eliminate their epinephrine and NaF response followed by reconstitution of these responses to membranes by a cytosolic preparation (Bonnafous *et al.*, 1979).

Several communications have documented a cytosolic component involved in the activation of adenylate cyclase by cholera toxin. A requirement for a cytosolic component and GTP in the activation and ADP-ribosylation of the G-protein was was demonstrated by Gill and co-workers (Gill and Meren, 1978; Gill, 1978; Gill and King, 1975; Enomoto and Gill, 1979). Partial purification and characterization of the macromolecular factor from pigeon (Enomoto and Gill, 1979) and horse erythrocytes (LeVine and Cuatrecasas, 1980a) failed to elucidate either the mechanistic role or the physiological identity of the factor, except to exclude the presence of calmodulin in the preparation. A possible connection between the requirement of the cytosolic factor for hormonal and cholera toxin-induced stimulation of adenylate cyclase has recently been shown. The purified cytosolic macromolecule from horse erythrocytes can, in a GTP-dependent fashion, potentiate the effect of glucagon and epinephrine on rat liver membranes (LeVine and Cuatrecasas, 1980a).

It remains a possibility that these cytosolic factors are enzymes capable of modifying membrane-bound components e.g., proteases, kinases, phosphatases, methylases, decarboxylases etc. However, the factors may equally well be involved more specifically in the coupling process.

8.6 PARTICIPATION OF THE CYTOSKELETON

The strongest evidence for the involvement of cytoskeletal elements in the process of hormonal activation of adenylate cyclase is the effect of various antimicrotubular agents on cAMP accumulation in whole cells first described by Rudolph *et al.* (1977). Colchicine, vinblastine and vincristine, which inhibit microtubular assembly, enhance catecholamine- and PGE_1 -induced cAMP accumulation (Kennedy and Insel, 1979; Kurokawa *et al.*, 1979; Gemsa *et al.*, 1977; Simantov and Sachs, 1978; Hagmann and Fishman, 1980; Dullis and Wilson, 1980; Zor *et al.*, 1978). Zor *et al.* (1978) have suggested that, in cultured rat Graffian follicles, different components of the cyto-skeleton may be involved in the response of adenylate cyclase to various agonists. Judged by their sensitivity to both colchicine and cytochalasin B (a microfilament-disrupting agent), choleragen and LH stimulation require microfilament integrity; PGE_1 requires microtubular function. In contrast, FSH stimulation seems to depend on both elements.

It has also been suggested that the cytoskeleton may be involved in hormonal desensitization. Wild type, desensitizing, leukemic cells which form normal caps in the presence of Concanavalin A respond to the alkaloids with a potentiated buildup of cAMP in response to hormones. In contrast, non-desensitizing clones of leukemic cells, which do not cap, are not affected by colchicine or vinblastine (Simantov and Sachs, 1978). These types of results have been interpreted as suggesting some involvement of the coupling components (which seem to be required for desensitization) with the cytoskeleton.

The effects of anti-cytoskeletal drugs require intact cells, whereas hormonal stimulation can be seen in broken cells or isolated membranes. It thus appears that the intact cytoskeleton may be involved in the *in vivo* modulation of hormonal sensitivity but not in the actual hormonal stimulation. The possible participation of some isolated components of the cytoskeleton in the mechanism of signal trans-duction will be considered below.

8.6.1 Association of adenylate cyclase components with the cytoskeleton

Although the hormonally sensitive adenylate cyclase is associated with the plasma membrane of the cell, the actual mode of association is largely unknown. While detergents seem to be required to remove adenylate cyclase from the membrane and maintain it in solution, recent observations indicate that the simple model of a hydrophobic association between enzyme and lipid matrix may be oversimplified. Studies from this laboratory, to be described later, have led us to differentiate between modes of association of the adenylate cyclase system with the bilayer or with elements of the cytoskeleton.

Proteins are associated with membranes in a number of different ways, varying from peripheral (i.e., having little or no hydrophobic contact with the bilayer, being mainly attached electrostatically, easily extractable without use of detergents;

the cytoskeleton is generally considered in this class) to integral (i.e. having a large hydrophobic interaction with the lipid matrix, generally only soluble in detergent).

It is apparent that these definitions are operational and reflect the skill and ingenuity of the investigator in 'solubilizing' a given protein. Some proteins require a specific sequence of extraction conditions for solubilization. If the extractions are not in the correct sequence, the result is often an insoluble precipitate dissociable only by harsh ionic detergent such as SDS or sometimes refractory to all but degradative conditions.

The approaches employed in studies we have carried out recently rely upon methodology developed to study the cytoskeleton of the red blood cell (Bennett and Stenbuck, 1979a and b; Tyler *et al.*, 1979). The red blood cell was chosen as a model because a number of potentially cytoskeleton-mediated events relevant to the adenylate cyclase system (e.g. hormonal stimulation and desensitization) occur in this well characterized and readily available tissue. Similar cytoskeletal proteins and functions appear to be present in other types of cells (Bennett, 1979). A detailed description of the techniques employed can be found in Bennett and Stenbuck (1979a and b) and Tyler *et al.*, (1979). A summary of the biochemical and physical evidence for cytoskeletal interactions from a variety of sources has recently been presented (Lux, 1979).

We have attempted to characterize the components of the adenylate cyclase complex in terms of their interactions with other proteins and/or the bilayer. This study has entailed the use of a variety of dissociating conditions and reconstitution with prepared cytoskeleton. The results are presented briefly in the following sections.

(a) *Evidence from differential extractions*

(i) *Adenylate cyclase.* Our results indicate that, at 0°C, nonionic detergents at concentrations sufficient to remove 70% of Band III solubilize less than 30% of the adenylate cyclase activity from rat erythrocyte ghosts. If the ghosts were repetitively extracted with 2% Triton X-100 at 0°C, 25 to 30% of the enzyme activity remained associated with the 'cytoskeleton'. Half of the cytoskeleton-associated activity could be released by treatment with low ionic strength buffers which disassemble the spectrin-actin network. These observations are now being pursued.

(ii) *β-Adrenergic receptors.* We have observed that, in the absence of ligand, β-receptor binding was lost after treatment with 0.03% Triton X-100 at 0°C, although no binding was detectable in the supernatant following extraction. However, in the presence of β-agonist or β-antagonist the receptor was lost in parallel with Band III, at much higher detergent concentrations. In these experiments the particulate fractions were washed extensively after extraction to completely remove any detergent before assaying for binding. In contrast to the results with adenylate cyclase, receptor binding was not altered by disassembly of the spectrin-actin network.

(iii) *Guanylnucleotide regulatory protein* (*G-protein*). The most definitive work on cytoskeletal interactions has been performed with the G-protein, the 42 000 mol. wt. protein ADP-ribosylated by cholera toxin. A significant proportion of the labeled band (30–40%) remained associated with the 'cytoskeleton' after repeated extractions with 2% Triton X-100 at $0°C$. The residual protein was not released by treatment with low ionic strength buffers at $37°C$. In contrast, about 30% of the protein could be eluted from untreated membranes with low ionic strength following which the remainder of the labelled protein could be eluted with nonionic detergents. The G-protein eluted by low ionic strength buffer migrated on gel filtration as a smaller molecular weight entity than the nonionic detergent extracted protein. This difference in apparent molecular size may be due to the detergent-solubilization of the G-protein with its 'anchor' or binding protein attached. Further investigation of this question is in progress.

These results imply a rather complicated association of the 42 000 mol. wt. factor with the membrane, probably involving cytoskeletal elements and possibly involving more than one pool of proteins. These separate pools, if they are conclusively demonstrated, may be related to both the distinct effects of guanylnucleotides on receptor affinity as well as regulation of the catalytic activity of adenylate cyclase.

8.6.2 Binding of solubilized components to the cytoskeleton

The reconstitution of a β-agonist-responsive adenylate cyclase into artificial phospholipid vesicles has been difficult to achieve. A variety of explanations could account for these observations, including the possibility that a vital link in the system is supplied by a cytoskeleton-associated element. This hypothesis lead to a series of experiments designed to reconstitute an hormonal response using the cytoskeleton as a matrix instead of a bilayer. From the results outlined below, to be described in detail in a subsequent publication, we feel the most promising approach will probably require a combination of lipid and cytoskeleton in a hybrid system yet to be developed.

(a) *Adenylate cyclase*

The catalytic subunit solubilized from rat erythrocyte or adipocyte membranes with nonionic detergent interacted with rat erythrocyte cytoskeleton in a time-, temperature-, and divalent cation-dependent manner. Binding was saturable and of high affinity. It could be abolished by trypsin digestion or heat treatment ($50°C$) of the cytoskeleton but was resistant to NEM. Preactivation of the enzyme with Gpp(NH)p, Gpp(NH)p + isoproterenol, or NaF was required for binding. The basal, non-stimulated, enzyme activity was not incorporated and could be recovered quantitatively in the supernatant, following removal of the cytoskeleton by centrifugation. These results suggest that there is some specific binding protein, for the catalytic unit or a tightly associated protein, present in the cytoskeleton.

(b) *β-Adrenergic receptors*

The most difficult reconstitution to perform remains that of the receptor. Using a variety of techniques we have not been able to detect any association of specific β-receptor binding activity with membranes, cytoskeletons or liposomes. These problems seem to be largely technical since it has recently been shown that solubilized β-receptors could be implanted into cells. Meanwhile, we are pursuing binding and reconstitution studies with a variety of other adenylate cyclase-linked and non-adenylate cyclase associated receptor systems whose physical properties may be more amenable to study.

(c) *Guanylnucleotide regulatory protein (G-protein)*

The binding to membranes of the 42 000 mol. wt. protein, ADP-ribosylated in the presence of cholera toxin, has been extensively examined with partially purified material. Previous studies (Farfel *et al.*, 1979), localizing this protein by indirect methods to the inner aspect to the erythrocyte membrane, have been confirmed in this laboratory. The binding is divalent cation-dependent ($Mn^{2+} > Mg^{2+} \approx Ca^{2+}$), only observed with inside-out vesicles (not ghosts) and is displaceable by unlabelled material. Both the labelled protein and the binding site on the membrane are trypsin- and heat-sensitive but are unaffected by NEM. Trypsin treatment of inside-out vesicles generated a soluble component capable of blocking binding of labelled protein. The results of this work were analogous to those obtained with spectrin and its binding protein, ankyrin. Binding of the ADP-ribosylated G-protein to cytoskeletons was similar but differed from that of adenylate cyclase in that it occurred only minimally at $0°C$ (as compared with $30°C$) and was absolutely dependent on the presence of $MgCl_2$.

Although work in this area is still preliminary, the possibility of examining the molecular architecture of the adenylate cyclase system is a stimulating one. The availability of antibodies to the recently purified G-protein and to the β-adrenergic receptor should aid greatly in this work.

8.7 ROLE OF LIPID IN ENZYME ACTIVITY

In recent years increasing attention has been paid to the role of the phospholipid matrix in modulating the activity of the adenylate cyclase system (Levey, 1973; Rethy *et al.*, 1972; Rubalcava and Rodbell, 1973; Limbird and Lefkowitz, 1975; Lad *et al.*, 1979). There is ample evidence to demonstrate that phospholipids play a role in modifying the activity of other membrane-bound enzymes (Coleman, 1973; Sandermann, 1978). Especially noteworthy in this regard is the data accumulated with the Ca^{2+}-transporting ATPase from sarcoplasmic reticulum (Warren *et al.*, 1975; Hesketh *et al.*, 1976), which is available in purified form.

In a recent publication we described the incorporation of adenylate cyclase, solubilized from rat brain, into artificial phospholipid vesicles (Hebdon *et al.*,1979).

We demonstrated that all of the enzyme was incorporated but that the activity expressed in the liposome was a function of the phospholipids present. We have been pursuing the question of the role of lipids in governing enzymatic activity, as have other laboratories. The data from many groups using a variety of techniques are currently not reconcilable with a unified picture. The following represents a summary of a representative cross section of the work. Although not all-inclusive, it is designed to show the breadth of investigation which is extant.

The techniques by which membrane phospholipid content is modified generally involve incubating cells or membranes with exogenous phospholipid (Martin and Macdonald, 1976; Bakardjieva *et al.,* 1979) or supplementing the medium in which auxotrophic cells are growing with phospholipid head groups or fatty acyl chains (Glaser *et al.,* 1974; Engelhard *et al.,* 1976 and 1978). This process, although it apparently maintains the integrity of the cell membrane, has some theoretical limitations. There may be regions of the membrane into which exogenously added lipid cannot partition or the lipid added may have other effects on the cell e.g. modifying metabolism or some other parameter.

In our laboratory we have adopted an alternative approach whereby we have solubilized the adenylate cyclase from membranes, dispersing the endogenous lipids, and assayed for activity on re-addition of defined phospholipids. This latter procedure, while overcoming the problem of partitioning and obviating any effect on metabolic status, has the drawback that hormonal stimulation cannot yet, in general, be resurrected. Gpp(NH)p and NaF stimulability, however, can be assessed in this preparation. The liposomal system also provides the technology to examine the effects of ion gradient and membrane potential upon enzyme activity.

8.7.1 Manipulation of the lipid composition

One accessible parameter with membranes is their bulk fluidity. Sinensky *et al.* (1979) studied a CHO–K_1 somatic cell mutant defective in the regulation of plasma membrane cholesterol content. Increasing the ordering of the membrane lipid by increasing the membrane cholesterol content led to an increase in basal adenylate cyclase activity. The absolute activity in the presence of PGE_1 or NaF also increased with increasing acyl chain ordering; however, this was apparently solely due to the elevation in basal activity. Thus the stimulation relative to basal activity was found to decrease in the presence of these agents. The authors' interpretation of enthalpic thermotropic changes led them to suggest that PGE_1, NaF and increased acyl chain ordering all activate adenylate cyclase by stabilizing a more active conformation of the enzyme. In apparent contrast with these results are data provided by Engelhard *et al.* (1976 and 1978) from studies with mouse LM cells. In these cells supplementation of the membranes with *cis* enoic fatty acids, which would be predicted to increase membrane fluidity, enhanced basal and NaF stimulated activities. PGE_1 stimulation was found to be directly proportional to the average number of double bonds present in the membrane fatty acyl chains (largely in phosphatidylcholine). They also found that in membranes derived from cells supplemented with ethanolamine, PGE_1-stimulated activity was essentially independent of the fatty acid supplement. This suggested that phosphatidylcholine and phosphatidylethanolamine may interact with adenylate cyclase in different ways.

Supplementation of LM cells with choline head group analogues also increased basal and PGE_1-stimulated activities whereas NaF stimulation was only significantly affected by L-2-amino-1-butanol supplementation, suggesting that the degree of substitution of the ethanolamine amino group may be a determinant of adenylate cyclase activity. They found that the K_m of the enzyme for ATP decreased with increasing substitution of the amino group. The authors found that PGE_1 stimulation did not vary systematically with membrane microviscosity and concluded that the variation in activity seen probably reflected direct or indirect interaction with phospholipid rather than a general sensitivity to bulk membrane viscosity.

Puchwein *et al.* (1974) found that addition of filipin (which complexes with cholesterol) reversibly uncoupled catecholamine activation of pigeon erythrocyte membrane adenylate cyclase. The GTP stimulation of adenylate cyclase was found to be even more sensitive to filipin addition, whereas binding of catecholamines to the receptor or basal enzymatic activity were little affected. Based on experiments with the hydrophobic fluorescent probe, perylene, the authors suggested that the structural order of the lipid matrix, rather than the microviscosity, is of prime importance for signal transfer from receptor to adenylate cyclase. Using the local anaesthetic benzyl alcohol, which increases membrane fluidity and consequently decreases the phase transition temperature, Houslay and colleagues (Dipple and Houslay, 1978) have shown an activation of adenylate cyclase. These results were interpreted as an effect of benzyl alcohol penetration of the inner shell (or 'annulus') of phospholipids surrounding the enzyme molecule, releasing some 'constraint' and consequently resulting in increased activity. The stimulation of the enzyme would be caused primarily by decreasing the energy of activation of the reaction.

There is some evidence to suggest that the outer lamella of the lipid bilayer may undergo a phase transition independently of the inner lamella. In support of this, Arrhenius plots of NaF- or Gpp(NH)p-stimulated adenylate cyclase activity did not show a break corresponding to the transition temperature of the outer leaflet (about 28.5°C for liver membranes) (Houslay *et al.*, 1976). However, the enzyme activity stimulated by glucagon did show an inflection at the transition temperature characteristic of the outer leaflet. Interestingly, the glucagon antagonist, des-his-glucagon also induced a sensitivity to the phase state of the outer leaflet. These results suggest that glucagon and des-his-glucagon but not NaF or Gpp(NH)p couple the receptor (or some transmembrane unit) to adenylate cyclase and thereby produce this sensitivity to the state of the outer leaflet.

Liver adenylate cyclase solubilized with the nonionic detergent Lubrol PX showed a break in the Arrhenius plot of enzymic activity at about 16°C, which is believed to represent a phase transition of Lubrol PX (m.p. 19°C) (Dipple and Houslay, 1978). This suggests that the uncoupled enzyme itself is sensitive to changes in the physical state of its environment. Similar detergent 'annulus' effects have been found with other membrane-bound enzymes.

A drastic manipulation of the phospholipid composition of the membrane by incorporating defined phospholipids into the membrane until they represented up to 60% of the total had no effect on the enzyme activity. This was interpreted as being due to a selection of certain phospholipids by the enzyme to constitute its 'annulus', excluding the added phospholipids. This dramatically illustrates one drawback of phospholipid supplementation as a way of modifying the microenvironment of a membrane enzyme. Recently the concept of a highly ordered ring of annular lipids has been called into question by some authors (Davoust *et al.*, 1980). This controversy is currently awaiting an unambiguous resolution.

The effects of lysophospholipids on the adenylate cyclase system has yielded conflicting data. It has been reported that, in neuroblastoma cells (Zwiller *et al.*, 1976) and murine 3T3 fibrobalsts (Shier *et al.*, 1976), lysophosphatidylcholine at 0.1–2 mM inhibits adenylate cyclase, whereas in rat liver plasma membranes (Houslay and Palmer, 1979) low concentrations of lysomyristoylphosphatidylcholine (0.001–0.01 mM) stimulated activity, while at higher concentrations (0.5–1 mM) inhibition was found. Lysooleoylphosphatidylcholine produced enzyme stimulation between 0.1–0.5 mM.

Glucagon stimulation of activity appeared to be inhibited at lower lysolecithin concentrations than needed to inhibit NaF stimulation. This inhibition was interpreted to be at the coupling level as a decrease in glucagon binding could not be demonstrated.

Recent experiments from the laboratory of Axelrod have raised the possibility that phospholipid methylation may be important in β-adrenergic stimulation of adenylate cyclase. They have described a β-agonist-induced methylation of phosphatidylethanolamine to phosphatidylcholine (Hirata *et al.*, 1979a). This process, which is dependent on S-adenosyl methionine, proceeds via phosphatidyl-N-monomethylethanolamine and phosphatidyl-N, N'-dimethylethanolamine (Hirata *et al.*, 1978; Hirata and Axelrod, 1978a). Axelrod and co-workers have correlated a decrease in membrane fluidity with the generation of phosphotidyl-N-monomethylethalolamine (Hirata and Axelrod, 1978b) and speculated that the decrease in virscosity would facilitate collision coupling (c.f., Rimon *et al.*, 1978) between hormone receptors and adenylate cyclase. They further demonstrated that S-adenosyl-homocysteine, a competitive inhibitor of S-adenosyl-methionin-dependent transmethylation reactions, reduced the stimulation of adenylate cyclase by isoproterenol.

It is interesting to note that the process of transmethylation has been implicated in many membrane phenomena including: expression of β-adrenergic receptors (Strittmatter *et al.*, 1979), chemotaxis (Hirata *et al.*, 1979b), lymphocyte mitogenesis (Hirata *et al.*, 1980), histamine release from mast cells (Hirata *et al.*, 1979a; Ishizaka *et al.*, 1980; Crews *et al.*, 1980) and prostaglandin synthesis (Hirata *et al.*, 1979c). However, recent work in this laboratory suggests that S-adenosyl-homocysteine and other inhibitors of transmethylation are, in addition, competitive inhibitors of cyclic nucleotide phosphodiesterase; interpretations of their actions should, therefore, take into consideration a possible effect, for example, of elevated cAMP levels.

8.7.2 Specific phospholipid requirements for adenylate cyclase activity

As alluded to previously, we have followed a somewhat different approach, solubilizing adenylate cyclase from membranes and assaying the activity in the presence of defined phospholipids. The principal source of enzyme we have used has been rat brain.

In initial studies adenylate cyclase solubilized with nonionic detergent became lipid-associated following incubation with exogenous phospholipids (Hebdon *et al.*, 1979). The enzyme appeared to associate integrally with the lipid matrix and was not released by high salt, or low salt/EDTA extraction, or by sonication. The association was blocked by increasing the viscosity of the bilayer through the addition of cholesterol. This ability to become lipid-associated suggests that either some region of the molecule is markedly hydrophobic (c.f., cytochrome b_5) (Ito and Sato, 1968; Spatz and Strittmatter, 1971; Robinson and Tanford, 1975) or, alternatively, that it is associating via some other membrane-bound protein.

Using this liposomal system preliminary studies were performed to investigate the effects of various phospholipids on enzymic activity. It was observed that maximal activity was found with phosphatidylcholine, addition of other lipids served only to decrease activity. There were two principal drawbacks to this study, first, nonionic detergent itself activates the brain enzyme (Johnson and Sutherland, 1973), and secondly, endogenous phospholipids solubilized with the enzyme complicate analysis of lipid effects. Definitive analysis of the effects of different lipids requires solubilization of the enzyme without nonionic detergent and also removal of endogenous phospholipids.

It was found that adenylate cyclase could be solubilized from rat brain with sodium deoxycholate (DOC). In this form enzymic activity was undetectable with either Mg^{2+} or Mn^{2+} in the absence of certain specific phospholipids or nonionic detergent. Maximal restoration of activity with both cations was found with phosphatidylcholine, sphingomyelin, phosphatidyl-*N*-monomethylethanolamine or lysophosphatidylcholine. Nonionic detergents, including Triton X-100 or Nonidet P40, were also able to restore full activity (see Table 8.2). Phosphatidylethanolamine and phosphatidyl-*N*, *N'*-dimethylethanolamine were capable of partially restoring activity whereas other lipids tested were completely without effect (see Table 8.2).

The stimulation by nonionic detergents admits two explanations; either the detergent is capable of directly causing activation, or it serves to allow the lipids present in the DOC extract to interact with the enzyme and cause activation. This ambiguity clearly demonstrated the need for a lipid-free enzyme preparation.

The adenylate cyclase activity and endogenous lipids in the DOC extract are easily separated by gel chromatography. The capacity of different phospholipids to reconstitute activity was again assayed but this time only sphingomyelin and

Table 8.2 Lipid requirement for adenylate cyclase activity

Sample	Maximally activating (80–100%)*	Partially activating (10–30%)*	Non-activating (0–5%)*
Whole brain extract	Sphingomyelin, lysolecithin, phosphatidylcholine, phosphatidyl-N-monomethyl-ethanolamine, nonionic detergent	Phosphatidylethanolamine, phosphatidyl-N,N'-dimethylethanolamine	Phosphatidylserine phosphatidylinositol, phosphatidic acid, phosphatidylglycerol, ceramide, ganglioside, free fatty acids, cholesterol, diglycerides, triglycerides
Delipidated enzyme	Sphingomyelin, lysolecithin	Phosphatidylcholine, nonionic detergent	All other lipids and lipid products tested

* Percentages are calculated relative to sphingomyelin.

lysolecithin were potent activators. Phosphatidylcholine and nonionic detergent were only partially effective; all other lipids were ineffective alone and inhibitory when added with sphingomyelin (Table 8.2). From Table 8.2 it can be seen that the lipid specificity becomes more apparent the less endogenous lipid is present. The ability of nonionic detergents and phosphatidylcholine to activate the enzyme is reduced upon delipidation. The residual activating capacity of these agents may be due to their genuine efficacy or may reflect the presence of remaining endogenous lipid.

The results we have obtained suggest that rat brain adenylate cyclase has a specific phospholipid requirement for activity, sphingomyelin and lysolecithin being efficacious whereas other lipids are not.

The phospholipid specificity for activity suggests a mechanism for modulating the enzyme. Modulation may involve the protein-catalyzed exchange of one phospholipid for another. If, for example, phosphatidylserine were exchanged for sphingomyelin, or phosphatidylcholine hydrolyzed to yield lysolecithin, one would predict that the activity of the enzyme would increase.

The reconstitution of activity by sphingomyelin is inhibited by low concentrations of phosphatidylethanolamine (maximum inhibition at < 25 mol %). On a more speculative note one may suggest that if the transmethylation seen by Axelrod and colleagues is central to hormone stimulation of adenylate cyclase, it may serve to remove tightly associated phosphatidylethanolamine from the enzyme and allow access of sphingomyelin which may induce activation of the enzyme.

There is one other interesting aspect of the interaction between enzyme and phospholipids which should be mentioned. Adenylate cyclase can be recovered in the cytosolic fraction from rat testis. This soluble enzyme is active only in the presence of Mn^{2+} and not Mg^{2+} (Braun and Dods, 1975). We have been able to incorporate this water-soluble enzyme into phosphatidylcholine vesicles, whereupon it becomes Mg^{2+} sensitive. These results raise the possibility that the metal ion specificity of the enzyme may be determined by its phospholipid environment.

The results presented above from many laboratories suggest that the phospholipid matrix of the membrane may exert controlling effects on the activity of adenylate cyclase. Whether these effects are of physiological relevance is the object of contiuing research and will have to await further experimental analysis.

8.8 CONCLUSION

The ultimate answer to the question: 'How do hormones stimulate adenylate cyclase?' will require the purification, in significant quantities, of all the components of the system and their functional integration into a phospholipid matrix with the full complement of cytoskeletal proteins. Only from this perspective will we finally be able to evaluate the relevance of the experimental results and approaches presented in this review.

REFERENCES

Bakardjieva, A., Galla, H.J. and Helmreich, E.J.M. (1979), *Biochemistry,* **18**, 3016–3023.

Bennett, V. and Stenbuck, P. (1979a), *J. biol. Chem.,* **254**, 2533–2541.

Bennett, V. and Stenbuck, P. (1979b), *Nature,* **280**, 468–473.

Bennett, V. (1979), *Nature,* **281**, 597–599.

Bennett, V. and Cuatrecasas, P. (1976), In: *The Specificity and Action of Animal, Bacterial and Plant Toxins, Receptors and Recognition* (Cuatrecasas, P., ed.), Series B, Vol. 1, pp. 1–66, Chapman and Hall, London.

Bird, S.J. and Maguire, M.E. (1978), *J. biol. Chem.,* **253**, 8826–8834.

Bonnafous, J.-C., Dornand, J. and Mani, J.-C. (1979), *FEBS Letters,* **99**, 152–156.

Bourne, H.R., Coffino, P. and Tomkins, G.M. (1975), *Science,* **187**, 750–752.

Bradham, L.S. (1977), *J. cyc. nucleot. Res.,* **3**, 119–128.

Braun, T. and Dods, R.F. (1975), *Proc. natn. Acad. Sci. U.S.A.,* **72**, 1097–1101.

Cassel, D., Eckstein, F., Lowe, M. and Selinger, Z. (1979), *J. biol. Chem.,* **254**, 9835–9838.

Cassel, D., Levkovitz, H. and Selinger, Z. (1977), *J. cyc. nucleot. Res.,* **3**, 393–406.

Cassel, D. and Pfeuffer, T. (1978), *Proc. natn. Acad. Sci. U.S.A.,* **75**, 2669–2673.

Cassel, D. and Selinger, Z. (1976), *Biochim. biophys. Acta,* **452**, 538–551.

Cassel, D. and Selinger, Z. (1977), *Proc. natn. Acad. Sci. U.S.A.,* **74**, 3307–3311.

Cassel, D. and Selinger, Z. (1978), *Proc. natn. Acad. Sci. U.S.A.,* **75**, 4155–4159.

Catt, K.J., Harwood, J.P., Aguilera, G. and Dufau, M.L. (1979), *Nature,* **280**, 109–116.

Coleman, R. (1973), *Biochim. biophys. Acta,* **300**, 1–30.

Crews, F.T., Morita, Y., Hirata, F. and Axelrod, J. (1980), *Biochem. biophys. Res. Comm.,* **93**, 42–49.

Cuatrecasas, P. (1974), *Ann. Rev. Biochem.,* **43**, 169–214.

Cuatrecasas, P., Jacobs, S. and Bennett, G.V. (1975), *Proc. natn. Acad. Sci. U.S.A.,* **72**, 1739–1743.

Daniel, V., Litwack, G. and Tomkins, G.M. (1973), *Proc. natn. Acad. Sci. U.S.A.,* **70**, 76–79.

Davoust, J., Bienvenue, A., Felliman, P. and Devaux, P.F. (1980), *Biochim. biophys. Acta.,* **596**, 28–42.

Dipple, I. and Houslay, M.D. (1978), *Biochem. J.,* **174**, 179–190.

Drummond, G.I., Sano, M. and Nambi, P. (1980), *Arch. biochem. biophys.,* **201**, 286–295.

Dufau, M.L., Hayashi, K., Sala, G., Bankal, A. and Catt, K.J. (1978), *Proc. natn. Acad. Sci. U.S.A.,* **75**, 4769–4773.

Dullis, B.H. and Wilson, I.B. (1980), *J. biol. Chem.,* **255**, 1043–1048.

Eimerl, S., Neufeld, G., Korner, M. and Schramm, M. (1980), *Proc. natn. Acad. Sci. U.S.A.,* **77**, 760–764.

Engelhard, V.H., Esko, J.D., Storm, D.R. and Glaser, M. (1976), *Proc. natn. Acad. Sci. U.S.A.,* **73**, 4482–4486.

Engelhard, V.H., Glaser, M. and Storm, D.R. (1978), *Biochemistry,* **17**, 3191–3200.

Enomoto, K. and Gill. D.M. (1979), *J. supramol. Struct.,* **10**, 51–60.

Farfel, Z., Kaslow, H.R. and Bourne, H.R. (1979), *Biochem. biophys. Res. Comm.*, **90**, 1237–1241.

Fox, C.F. and Das, M. (1979), *J. supramol. Structure*, **10**, 199–214.

Ganguly, U. and Greenough, III, W.B. (1975), *Proc. natn. Acad. Sci. U.S.A.*, **72**, 3561–3564.

Gemsa, D., Steggemann, L., Till, G. and Resch, K. (1977), *J. Immunol.*, **119**, 524–529.

Gill, D.M. (1976), *J. Infectious Dis.*, **133**, S55–S63.

Gill, D.M. (1977), *Adv. cyc. nucleot. Res.*, **8**, 85–118.

Gill, D.M. and King, C.A. (1975), *J. biol. Chem.*, **250**, 6424–6432.

Gill, D.M. and Meren, R. (1978), *Proc. natn. Acad. Sci. U.S.A.*, **75**, 3050–3054.

Glaser, M., Ferguson, K.A. and Vagelos, P.R. (1974), *Proc. natn. Acad. Sci. U.S.A.*, **71**, 4072–4076.

Goldstein, J.L., Anderson, R.G.W. and Brown, M.S. (1979), *Nature*, **279**, 679–685.

Grollman, E.F., Lee, G., Ambesi-Impiombato, F.S., Meldolesi, M.F., Aloj, S.M., Coon, H.G., Kaback, H.R. and Kohn, L.D. (1977), *Proc. natn. Acad. Sci. U.S.A.*, **74**, 2352–2356.

Guillon, G., Couraud, P.O. and Roy, C. (1979), *Biochem. biophys. Res. Comm.*, **87**, 855–861.

Haga, T., Haga, K. and Gilman, A.G. (1977a), *J. biol. Chem.*, **251**, 5776–5782.

Haga, T., Ross, E.M., Anderson, H.J. and Gilman, A.G. (1977b), *Proc. natn. Acad. Sci. U.S.A.*, **74**, 2016–2020.

Hagmann, J. and Fishman, P.H. (1980), *J. biol. Chem.*, **255**, 2659–2662.

Hebdon, G.M., LeVine, III, H., Minard, R.B., Sahyoun, N.E., Schmitges, C.J. and Cuatrecasas, P. (1979), *J. biol. Chem.*, **254**, 10459–10465.

Hebdon, G.M., LeVine, III, H., Sahyoun, N.E., Schmitges, C.J. and Cuatrecasas, P. (1978), *Proc. natn. Acad. Sci. U.S.A.*, **75**, 3693–3697.

Hesketh, T.R., Smith, G.A., Houslay, M.D., McGill, K.A., Birdsall, N.J.M., Metcalfe, J.C. and Warren, G.B. (1976), *Biochemistry*, **15**, 4145–4151.

Hirata, F. and Axelrod, J. (1978a), *Nature*, **275**, 219–220.

Hirata, F. and Axelrod, J. (1978b), *Proc. natn. Acad. Sci. U.S.A.*, **75**, 2348–2352.

Hirata, F., Axelrod, J. and Crews, F.T. (1979a), *Proc. natn. Acad. Sci. U.S.A.*, **76**, 4813–4816.

Hirata, F., Corcoran, B.A., Venkatasubramanian, K., Schiffman, E. and Axelrod, J. (1979b), *Proc. natn. Acad. Sci. U.S.A.*, **76**, 368–372.

Hirata, F., Corcoran, B.A., Venkatasubramanian, K., Schiffman, E. and Axelrod, J. (1979c), *Proc. natn. Acad. Sci. U.S.A.*, **76**, 2640–2643.

Hirata, F., Strittmatter, W.J. and Axelrod, J. (1979a), *Proc. natn. Acad. Sci. U.S.A.*, **76**, 368–372.

Hirata, F., Toyoshima, S., Axelrod, J. and Waxdal, M.J. (1980), *Proc. natn. Acad. Sci. U.S.A.*, **77**, 862–865.

Hirata, F., Viveros, O.H., Diliberto, E.M., Jr. and Axelrod, J. (1978), *Proc. natn. Acad. Sci. U.S.A.*, **75**, 1718–1721.

Hoffman, F.M. (1979a), *J. biol. Chem.*, **254**, 255–258.

Hoffman, F.M. (1979b), *Biochem. biophys. Res. Comm.*, **86**, 988–994.

Houslay, M.D., Ellory, J.C., Smith, G.A., Hesketh, T.R., Stein, J.M., Warren, G.B. and Metcalfe, J.C. (1977), *Biochim. biophys. Acta*, **467**, 208–219.

Houslay, M.D., Hesketh, T.R., Smith, G.A., Warren, G.B. and Metcalfe, J.C. (1976), *Biochim. biophys. Acta,* **436**, 495–504.

Houslay, M.D. and Palmer, R.W. (1979), *Biochem. J.,* **178**, 217–221.

Howlett, A.C., Sternweis, P.C., Macik, B.A., Van Arsdale, P.M. and Gilman, A.G. (1979), *J. biol. Chem.,* **254**, 2287–2295.

Insel, P.A., Bourne, H.R., Coffino, P. and Tomkins, G.M. (1975), *Science,* **190**, 896–898.

Insel, P.A., Maguire, M.E., Gilman, A.G., Bourne, H.R., Coffino, P. and Melmon, K.L. (1976), *Mol. Pharmacol.,* **12**, 1062–1068.

Ishizaka, T., Hirata, F., Ishizaka, K. and Axelrod, J. (1980), *Proc. natn. Acad. Sci. U.S.A.,* **77**, 1903–1906.

Ito, A. and Sato, R. (1968), *J. Biol. Chem.,* **243**, 4922–4923.

Jacobs, S. and Cuatrecasas, P. (1976), *J. cyc. nucleot. Res.,* **2**, 205–223.

Johnson, G.L., Bourne, H.R., Gleason, M.K., Coffino, P., Insel, P.A. and Melmon, K.L. (1979), *Mol. Pharmacol.,* **15**, 16–27.

Johnson, G.L., Kaslow, H.R. and Bourne, H.R. (1978b), *J. biol. Chem.,* **253**, 7120–7123.

Johnson, G.L., Kaslow, H.R. and Bourne, H.R. (1978a), *Proc. natn. Acad. Sci. U.S.A.,* **75**, 3113–3117.

Johnson, R.A. and Sutherland, E.W. (1973), *J. biol. Chem.,* **248**, 5114–5121.

Kaslow, H.R., Farfel, Z., Johnson, G.L. and Bourne, H.R. (1979), *Mol. Pharmacol.,* **15**, 472–483.

Katz, M.S., Kelly, T.M., Pineyro, M.A. and Gregerman, R.I. (1978), *J. cyc. nucleot. Res.,* **5**, 389–407.

Kennedy, M.S. and Insel, P.A. (1979), *Mol. Pharmacol.,* **16**, 215–223.

Kurokawa, T., Kurokawa, M. and Ishibashi, S. (1979), *Biochim. biophys. Acta,* **583**, 467–473.

Laburthe, M., Rosselin, G., Rousset, M., Zweibaum, A., Korner, M., Selinger, Z. and Schramm, M. (1979), *FEBS Letters,* **98**, 41–43.

Lad, P.M., Nielsen, T.B., Preston, M.S. and Rodbell, M. (1980), *J. biol. Chem.,* **255**, 988–995.

Lad, P.M., Preston, S., Welton, A.F., Nielsen, T.B. and Rodbell, M. (1979), *Biochim. biophys. Acta,* **551**, 368–381.

Leavitt, W.W. and Clark, J.H., eds. (1979), *Steroid Hormone Receptor Systems,* Plenum Press, New York.

Levey, G.S. (1973), *Rec. Progr. Hormone Res.,* **29**, 361–382.

LeVine, III, H. and Cuatrecasas, P. (1980a), (manuscript submitted).

LeVine, III, H. and Cuatrecasas, P. (1980b) *J. Pharmacol. Therap.* In press.

Limbird, L.E., Hickey, A.R. and Lefkowitz, R.J. (1979), *J. cyc. nucleot. Res.,* **5**, 251–259.

Limbird, L.E. and Lefkowitz, R.J. (1975), *Mol. Pharmacol.,* **12**, 559–567.

Limbird, L.E. and Lefkowitz, R.J. (1977), *J. biol. Chem.,* **252**, 799–802.

Londos, C., Salomon, Y., Lin, M.C., Harwood, J.P., Schramm, M., Wolfe, J. and Rodbell, M. (1977), *Proc. natn. Acad. Sci. U.S.A.,* **71**, 3087–3090.

Lux, S.E. (1979), *Nature,* **281**, 426–429.

Maguire, M.E. and Erdos, J.J. (1978), *J. biol. Chem.,* **253**, 6633–6636.

Maguire, M.E. and Erdos, J.J. (1980), *J. biol. Chem.*, **255**, 1030–1035.
Maguire, M.E., Ross, E.M. and Gilman, A.G. (1977), *Adv. cyc. nucleot. Res.*, **8**, 1–83.
Martin, B.R., Stein, J.M., Kennedy, E.L., Doberska, C.A. and Metcalfe, J.C. (1979), *Biochem. J.*, **184**, 253–260.
Martin, F.J. and Macdonald, R.C. (1976), *J. cell Biol.*, **70**, 515–526.
Monneron, A. and d'Alayer, J. (1980), *FEBS Letters*, **109**, 75–80.
Naya-Vigne, J., Johnson, G.L., Bourne, H.R. and Coffino, P. (1978), *Nature*, **272**, 720–722.
Orly, J. and Schramm, M. (1976), *Proc. natn. Acad. Sci. U.S.A.*, **73**, 4410–4414.
Pecker, F. and Hanoune, J. (1977a), *FEBS Letters*, **83**, 93–98.
Pecker, F. and Hanoune, J. (1977b), *J. biol. Chem.*, **252**, 2784–2786.
Pfeuffer, T. (1977), *J. biol. Chem.*, **252**, 7224–7234.
Pfeuffer, T. (1979), *FEBS Letters*, **101**, 85–89.
Pfeuffer, T. and Helmreich, E.J.M. (1975), *J. biol. Chem.*, **250**, 867–876.
Puchwein, G., Pfeuffer, T. and Helmreich, E.J.M. (1974), *J. biol. Chem.*, **249**, 3232–3240.
Renart, M.F., Ayanoglu, G., Mansour, J.M. and Mansour, T.E. (1979), *Biochem. biophys. Res. Comm.*, **89**, 1146–1153.
Rethy, A., Tomasi, V., Trevisani, A. and Barnobei, O. (1972), *Biochim. biophys. Acta*, **290**, 58–69.
Rimon, G., Hanski, E., Braun, S. and Levitzki, A. (1978), *Nature*, **276**, 396–399.
Robinson, N.C. and Tanford, C. (1975), *Biochemistry*, **14**, 369–378.
Rodbell, M. (1980), *Nature*, **284**, 17–22.
Rodbell, M., Birnbaumer, L., Pohl, S.L. and Krans, H.M.J. (1971), *J. biol. Chem.*, **246**, 1877–1882.
Rodbell, M., Lin, M.C., Salomon, Y., Londos, C., Harwood, J.P., Martin, B.R., Rendell, M. and Berman, M. (1975), *Adv. cyc. nucleot. Res.*, **5**, 3–29.
Ross, E.M. and Gilman, A.G. (1977a), *Proc. natn. Acad. Sci. U.S.A.*, **74**, 3715–3719.
Ross, E.M. and Gilman, A.G. (1977b), *J. biol. Chem.*, **252**, 6966–6969.
Ross, E.M., Howlett, A.C., Ferguson, K.M. and Gilman, A.G. (1978), *J. biol. Chem.*, **253**, 6401–6412.
Ross, E.M., Maguire, M.E., Sturgill, T.W., Biltonen, R.L. and Gilman, A.G. (1977), *J. biol. Chem.*, **252**, 5761–5775.
Rubalcava, B. and Rodbell, M. (1973), *J. biol. Chem.*, **248**, 3831–3837.
Rudolph, S.A., Greengard, P. and Malawista, S.E. (1977), *Proc. natn. Acad. Sci. U.S.A.*, **74**, 3404–3408.
Sahyoun, N.E., Schmitges, C.J., LeVine, III, H. and Cuatrecasas, P. (1977), *Life Sciences*, **21**, 1857–1864.
Sandermann, H. (1978), *Biochim. biophys. Acta*, **515**, 209–237.
Sanders, R.B., Thompson, W.J. and Robison, G.A. (1977), *Biochim. biophys. Acta*, **498**, 10–20.
Sano, K., Katsuda, K. and Maeno, H. (1978), *Neurochem. Res.*, **3**, 674.
Sano, K., Nishikori, K., Noshiro, O. and Maeno, H. (1979), *Arch. biochem. Biophys.*, **197**, 285–293.
Schlegel, W., Kempner, E.S. and Rodbell, M. (1979), *J. biol. Chem.*, **254**, 5168–5176.
Schramm, M. (1979), *Proc. natn. Acad. Sci. U.S.A.*, **76**, 1174–1178.

Schramm, M., Orly, J., Eimerl, S. and Korner, M. (1977), *Nature,* **268**, 310–313.

Schramm, M. and Rodbell, M. (1975), *J. biol. Chem.,* **250**, 2232–2237.

Schulster, D., Orly, J., Seidel, G. and Schramm, M. (1978), *J. biol. Chem.,* **253**, 1201–1206.

Schwarzmeier, J.D. and Gilman, A.G. (1977), *J. cyc. nucleot. Res.,* **3**, 227–238.

Shier, W.T., Baldwin, J.H., Nilsen-Hamilton, M., Hamilton, R.T. and Thanassi, N.M. (1976), *Proc. natn. Acad. Sci. U.S.A.,* **73**, 1586–1590.

Simantov, R. and Sachs, L. (1978), *FEBS Letters,* **90**, 69–75.

Sinensky, M., Minneman, P. and Molinoff, P.B. (1979), *J. biol. Chem.,* **254**, 9135–9141,

Spatz, L. and Strittmatter, P. (1971), *Proc. natn. Acad. Sci. U.S.A.,* **68**, 1042–1046.

Sternweis, P.C. and Gilman, A.G. (1979), *J. biol. Chem.,* **254**, 3333–3340.

Strittmatter, W.J., Hirata, F. and Axelrod, J. (1979), *Science,* **204**, 1205–1207.

Toscano, Jr., W.A., Wescott, K.R., LaPorte, D.C. and Storm, D.R. (1979), *Proc. natn. Acad. Sci. U.S.A.,* **76**, 5582–5586.

Tyler, J.M., Hargreaves, W.R. and Branton, D. (1979), *Proc. natn. Acad. Sci. U.S.A.,* **76**, 5192–5196.

Vaughan, M. and Moss, J. (1978), *J. supramol. Struct.,* **8**, 473–488.

Wagner, H.R. and Davis, J.N. (1979), *Proc. natn. Acad. Sci. U.S.A.,* **76**, 2057–2061.

Warren, G.B., Houslay, M.D., Metcalfe, J.C. and Birdsall, N.J.M. (1975), *Nature,* **225**, 684–687.

Williams, L.T., Mullikin, D. and Lefkowitz, R.J. (1978), *J. biol. Chem.,* **253**, 2984–2989.

Zor, U., Strulovici, B. and Lindner, H.R. (1978), *Biochem. biophys. Res. Comm.,* **80**, 983–992.

Zwiller, J., Ciesielski-Treska, J. and Mandel, P. (1976), *FEBS Letters,* **69**, 286–290.

Index